Alfred Wilks Drayson

On the Cause, Date and Duration of the Last Glacial Epoch of Geology and the Probable Antiquity of Man

Alfred Wilks Drayson

On the Cause, Date and Duration of the Last Glacial Epoch of Geology and the Probable Antiquity of Man

ISBN/EAN: 9783741141256

Manufactured in Europe, USA, Canada, Australia, Japa

Cover: Foto ©berggeist007 / pixelio.de

Manufactured and distributed by brebook publishing software (www.brebook.com)

Alfred Wilks Drayson

On the Cause, Date and Duration of the Last Glacial Epoch of Geology and the Probable Antiquity of Man

SUMMER IN NORTHERN HEMISPHERE.

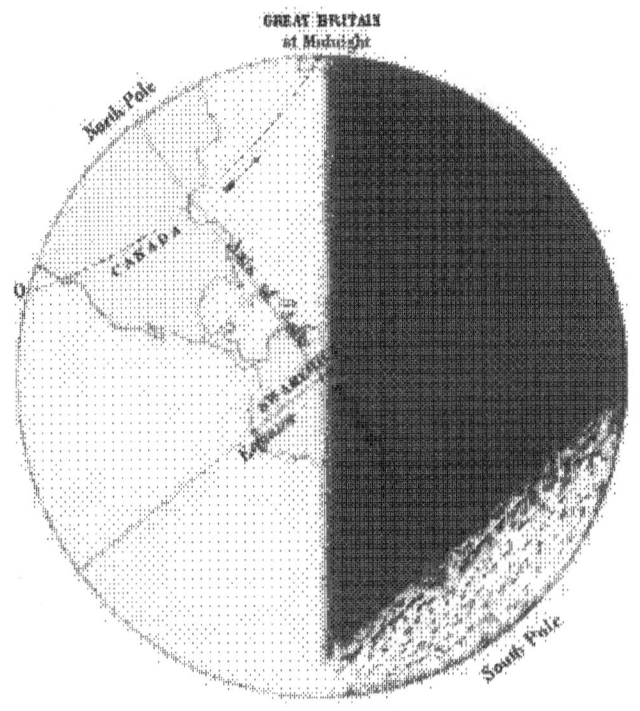

Diagram showing the extent of perpetual day in Summer during the Glacial Epoch 12,600 B.C.

Great Britain would at that date have had perpetual day, being carried round the circle shown by the arrow every 24 hours.

Parts of Canada also, would have had perpetual day.

O Shows position of Great Britain at Midday. Sun's meridian altitude then 74° 15' instead of 62° as at present.

WINTER IN NORTHERN HEMISPHERE.

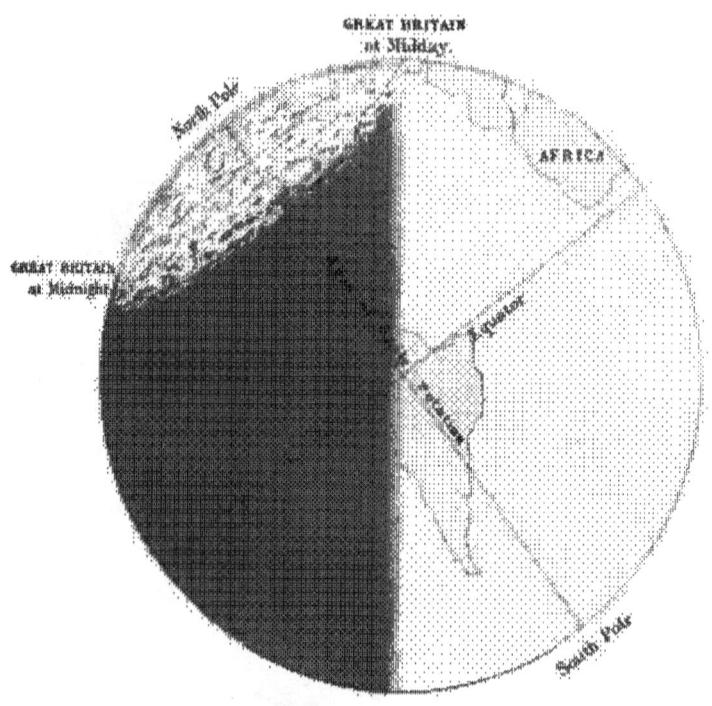

Diagram showing the extent of the Arctic circle during the height of the Glacial Epoch 12,600 B. C.

Showing how perpetual darkness would then prevail during several weeks in Great Britain thus causing icebergs and snow in immense quantities.

The arrow shows the course of the diurnal rotation.

ON THE

Cause, Date, and Duration

OF THE

LAST GLACIAL EPOCH OF GEOLOGY,

AND THE PROBABLE ANTIQUITY OF MAN.

WITH AN INVESTIGATION AND DESCRIPTION OF A NEW MOVEMENT OF THE EARTH.

BY

LIEUT.-COL. DRAYSON, R.A. F.R.A.S.

AUTHOR OF COMMON SIGHTS IN THE HEAVENS, PRACTICAL MILITARY SURVEYING, ETC.

LONDON:
CHAPMAN AND HALL, 193 PICCADILLY.
1873.

PREFACE.

In the following pages will be found a description of the most important facts connected with what is called the 'Glacial Epoch' of Geology. Having given an account of this singular period in the earth's history, we give a brief description of the theories which have at various times been put forward and have been supposed to account for the facts. These theories we briefly refer to, and endeavour to call attention to their inefficiency to explain the effects of which they are the conjectured causes.

We then examine the three principal movements of the earth, and point out how one of these, as at present interpreted, presents a geometrical impossibility which renders the interpretation untenable.

An examination of this movement of the earth reveals the fact that it is one which fully explains the recorded observations of astronomy with minute accuracy, which the present accepted movement cannot explain, and also gives a full and complete explanation of the Glacial Epoch of Geology, and not only affords an explanation of the effects, but gives the date and duration of this mysterious period.

Having endeavoured to obtain a fair criticism on the views put forward relative to this movement of the earth, a few of the objections that have been urged are quoted and replied to.

A brief account is given of some of the different results which will follow this movement of the earth; but, as the astronomical portion will be given in another work, these results are merely mentioned in as few words as possible. Several years having been devoted to the working out of all parts of this problem and this new movement of the earth, the following pages must be considered as only a mere outline sketch of some of the most prominent effects.

CONTENTS.

CHAPTER I.
BRIEF DESCRIPTION OF THE GLACIAL EPOCH, AND ITS CONJECTURED CAUSES 1

CHAPTER II.
EXTRACTS FROM GEOLOGICAL WORKS RELATIVE TO THE GLACIAL EPOCH 8

CHAPTER III.
CONSIDERATION OF THE CONJECTURED CAUSES SUPPOSED TO ACCOUNT FOR THE GLACIAL EPOCH . . . 79

CHAPTER IV.
THE THREE PRINCIPAL MOVEMENTS OF THE EARTH DESCRIBED AND EXAMINED 96

CHAPTER V.
EXAMINATION OF THE THIRD MOVEMENT OF THE EARTH, AND THE TRUE COURSE TRACED BY THE EARTH'S AXIS 115

CHAPTER VI.
THE CLIMATE OF THE EARTH DURING 31,840 YEARS . 144

CONTENTS.

CHAPTER VII.
OBJECTIONS RAISED AND CONSIDERED . . . 179

CHAPTER VIII.
OBJECTIONS CONTINUED . . . 215

CHAPTER IX.
CONSIDERATIONS RELATIVE TO GEOLOGICAL TIME AND THE ANTIQUITY OF MAN 242

CHAPTER X.
OUTLINES OF SOME OF THE RESULTS OF THE CURVE TRACED BY THE EARTH'S AXIS . . . 259

CHAPTER XI.
CONCLUDING REMARKS 270

THE
LAST GLACIAL EPOCH OF GEOLOGY.

CHAPTER I.

BRIEF DESCRIPTION OF THE GLACIAL EPOCH AND ITS CONJECTURAL CAUSES.

AMONG the various conditions to which our earth has been subjected, there are none which present more singular or interesting phases than that known as the 'Glacial Epoch' or 'Boulder Period.' There are two special reasons why this remarkable period should attract our attention, and these are: that the evidences of this epoch even now lie on the earth's surface as they were left by the forces which produced them; also, the boulder period seems to be the last great change through which our planet has passed previous to its arriving at the present condition.

Thus, although there must be difficulties to surmount in tracing out the causes which produced the various changes exhibited by the crust of the earth, yet it would seem that these difficulties ought to be less, when we devote our attention to investigating a problem the data of which are exhibited on the actual surface of the earth, and extend over the northern and southern portions of each hemisphere, and in almost every instance seem to have been unaltered by the passage of time which has elapsed since their occurrence.

What these evidences are may be briefly stated as follows:

It appears that the whole northern and southern hemispheres were subjected to a cold climate of such intensity as to cause an arctic climate to prevail down to about 45° or 50° of latitude. The result was, that icebergs large enough to carry immense blocks of stones, and to transport them on to distant localities, were formed in England, Wales, and various parts of Europe and America, where such icebergs are never now formed; also there is evidence that glaciers prevailed in the same districts. These and other corresponding facts have led most reasoners to conclude, that from some unexplained cause, the surface of the earth, at least as far south as 50° of latitude, was covered by ice, and that it was the moulding effect of this ice which is now exhibited on the rocks and soil, to give evidence of its former power.

The period during which these remarkable events occurred is that known as the 'glacial,' and it has been for upwards of thirty years a subject of deep interest among inquirers, who have endeavoured, and whose labours have been devoted, to find a cause for the effects, or to obtain some clue to the date and duration of this remarkable period of the earth's history.

Hitherto the causes assigned have on close investigation been found either inadequate, or mere arbitrary assumptions; or have been supposed the effect of some catastrophe of Nature; and as such are unsatisfactory to those who consider that Nature acts uniformly, and not by a series of convulsions. Mathematicians have also rejected some of the assumptions supposed to explain the facts, because it has been found that these were opposed to the best-known laws of mechanics. In every case, however, the assumed or suggested causes have left one interesting subject in entire obscurity, for they have never dealt upon sound principles with the date or duration of the Boulder period. Whether this occurred twenty thousand or twenty million years ago, is a mere guess. Whether the period lasted ten thousand or ten million years, is also a mere matter of guess, and so it must remain, unless we can discover a cause for the effects; for we know not whether this cause acted with great force and quickly, or with weakness and slowly.

It has been urged as an example, that a pile may be driven into the ground by an endless number of weak blows extending over an immense period of time, or by a few rapid and powerful blows requiring only a short time. This argument may with slight modification be applied to geological time, and hence the glacial period may, to some extent, be uncertain in its duration; for if the forces were weak, their duration must have been long; if they were powerful, their duration need only have been brief to produce all the effects.

It will probably be evident that if we find a cause which would produce the effects of the glacial period, and that this cause is even now at work, we have merely to trace back the conditions, and find at what date they were such as to produce the known effects. Also by continuing to trace out the conditions, we can find when the effects ought to have ceased. Thus it may be stated, that the cause, date, and duration of the glacial epoch is one and the same problem. If we can find the cause, we may probably also ascertain the date and duration, or at least a very close approximation for them; and although this glacial period is only one small day, as it were, in the world's history, still if we can find a cause for this day being different from our present day, we make one advance towards the solution of a problem, the vastness of which has hitherto entirely defeated inquiry.

Confining our investigations, therefore, entirely to the Boulder period, we will extract from the works of various geologists such descriptions as will enable the reader to form an opinion as regards the facts which are known and recognised. Previous to

giving these extracts, we will briefly refer to the
supposed causes which have been suggested as likely
to have produced these effects. These supposed
causes cannot be termed more than theories, for they
themselves require proof, and in almost all cases
occupy the position of a '*possible*' cause, which, if it
happened, might explain some of the observed facts;
but not having any evidence to show that this cause
ever occurred, it is little more than a baseless theory.

    1st. One assumed cause is, that the Earth, travel-
ling with the solar system through space,
sometimes passes through warm, sometimes
through cold regions; and thus when it
passed through a cold region the glacial
epoch occurred.

    2d. That the shape of the Earth's orbit has al-
tered, and that formerly the Earth, during
one year, varied its distance from the Sun
much more than at present, and possibly
this annual alteration might occasion the
variation of climate.

    3d. That considerable alterations have occurred
in the elevation and depression of land, and
that the relative position of land and sea
causes great climatic changes; hence probably during the glacial period the land and
sea were so distributed as to cause a cold
climate over the whole northern hemisphere.

    4th. That possibly the Earth's axis, instead of
being located where it now is, in the Earth,

had a different location, the North and South Poles being situated in other positions than those they now occupy. If this occurred, of course great climatic changes would have resulted, which it is suggested may have occasioned the climate of the glacial epoch.

These and various other theories have been put forward to account for the facts of the Boulder period. Whilst granting their ingenuity in some respects, it will be necessary to analyse them carefully, in order to show either their insufficiency to explain that which is known to have occurred, or the great improbability of their occurrence. This investigation will be made after we have collected evidence relative to the various phenomena observed by geologists relative to the glacial epoch. In addition to the great interest which must attach to the investigation of such a problem as the cause or causes which produced the glacial epoch, and the inquiry as to its length and the interval of time which has elapsed since its occurrence,—there is another subject intimately connected with this period, and which has also occupied the attention of inquirers of all nations for many years, viz. the Antiquity of Man.

It seems to be now generally granted that human beings were on earth during the glacial epoch; thus, if we can ascertain at what date this epoch occurred, we can at once fix with some pretence to accuracy the length of time during which man has been a denizen of our planet. We may not be able to limit the time, and say it was so many years, and no longer; but if we can demonstrate that, say, twenty thousand years ago, such conditions prevailed as would of themselves have produced all the known effects of the great Boulder period, then we can state that if the evidence is undeniable that man did exist on earth during that condition, then he has at least been twenty thousand years on earth.

In fact this problem resolves itself into that of converting geological periods into astronomical time, a subject which some years ago occupied the attention of most geologists and mathematicians, but which it appears was one which, after inquiry, remained much in the same state that it was before inquiry; for no advance was made towards its solution. Thus whilst we devote our attention to the investigation of a problem in Geology, we also are directing our inquiries to the solution of another subject, equally interesting, viz. the Antiquity of Man. Hence we may fairly conclude, that dry as most of the details of a scientific investigation must necessarily prove to some readers, yet the importance of the subjects treated of, and the entire novelty of the explanations brought to notice, will, we trust, cause some interest to be taken in the following pages.

CHAPTER II.

EXTRACTS FROM GEOLOGICAL WORKS RELATIVE TO THE GLACIAL EPOCH.

Extracted from 'Physical Geology of Great Britain,' by Professor Ramsey.

BUT I must now describe a remarkable episode in the latest Tertiary (or, as some authors call them, the Post Tertiary) times, known as the 'Glacial Epoch,' said to be altogether of later date than the Cromer 'Forest Bed,' and certainly of earlier date than some of the patches just alluded to. This formation has left its traces universally over the whole northern half of the northern, and also over a large portion of the southern, hemisphere; and I hope I shall be able to describe the history of that period, as it affects the scenery of Britain, with something like tolerably accurate detail. Before doing so, however, I must lead you into Switzerland, and show you what kind of effect is being produced there by the ice of the present day; and afterwards into Greenland, and show you what takes place there; and then, by the knowledge thus gained, I shall be able to bring you back into our own country, and explain what took place here in that icy episode which is so far distant in time, but which, by comparison with the more ancient periods, almost approaches our own day.

Now the first thing I have to do is to describe what a glacier is. In any large and good map of Switzerland, you will see certain white patches here and there on the higher mountain ranges of the Alps. The highest mountain in the Alps, Mont Blanc, rises more than 15,000 feet above the sea; and there are other mountains in this great chain which approach that height, ranging from 12,000 to 15,000 feet. The mean limit of perpetual snow upon the Alps is 8500 feet above the level of the sea. Above that line, speaking generally, the country is mostly covered with snow; and in the higher regions it gathers on the mountain slopes and in the larger recesses, and by force of gravity it presses downwards into the main valleys, where, chiefly in consequence of the immense pressure exerted by the weight and movement of this accumulated mass, the snow year after year is converted into ice.

Without entering on details, it is enough if I now state that this is proved by well-considered observations, made by the best observers of the icy phenomena of the Alps. Still accumulating year upon year, by degrees this ice slides down the valleys, and is often protruded in a great tongue far below the limits of perpetual snow; for some glaciers descend as low as from 3000 or 4000 feet or thereabouts above the level of the sea, whereas the limit of perpetual snow is 8500 feet.

Now I will not enter into all the details of the structure of glaciers, because that will not help us in the special investigation we have now in view; but I will describe to you what are the effects produced by a glacier in the country over which it slides, and various other glacier-phenomena affecting the scenery of the Alps, and therefore affecting the scenery of our own country in past times, when glaciers existed here, and still affecting it in the relics they have left.

A glacier slides more or less rapidly, according to the mass of ice that fills the valley, and also according to the greater or less inclination of the slope; for in this respect it behaves very like a river. If you have a vast body of water like the Mississippi flowing down a broad valley, although the slope of the valley may be very gentle, still the river flows with great rapidity, in consequence of the greatness of the body of water. So if you have a mass of ice, which represents the snow-drainage of a large tract of country *covered with perpetual snow*, then the *glacier flows with a rapidity proportionate to the mass of ice;* and that rate of progress is modified, increased, or diminished in accordance with the fall and width of the valley; so that when it is steep, the glacier flows comparatively fast; and when the angle at which the valley slopes is small, it flows with comparative slowness.

All glaciers are traversed by cracks, which are termed 'crevasses.' Now the mountain-peaks that rise above the surface of a glacier in some cases are so steep, that the snow refuses to lie upon them, even when they may happen to be above the limits of the average line of perpetual snow; so that masses of rock are always being severed by atmospheric disintegration, and, falling from the slopes, they find a temporary resting-place on the surface of the ice at the margin of the glacier, and as it were float upon its surface in long and continuous lines; for the motion of a glacier is so slow, that the quantity of stones that fall upon its surface is sufficiently numerous to keep up a continuous line of blocks, earth and gravel, often of great width.

These stones, when two glaciers combine to form one great stream of ice—as in the glacier of the Aar—at a certain point meet and form one grand line running down the centre of the glacier. These are termed 'moraines;' and at length all of this material that has not fallen into crevasses floats on as it were to the end of the glacier, and is shot into the valley at the end of the ice-stream, frequently forming large mounds, known as 'terminal moraines.' Beneath every glacier water is constantly flowing, caused by the melting of the ice both below and on the surface of the glacier, and also, in some cases to a less degree, by springs that rise in the rocks below the ice. In the various parts of glaciers where crevasses are not numerous, you frequently find large brooks, so wide that you cannot leap across them, and you may have to walk half a mile before you can find a passage. But in all the glaciers that I have seen, long before you reach their lower end all the surface-water has found its way to the bottom of the ice. The water therefore that runs from the end of the glacier very often emerges from an ice-cavern as a large, ready-made, muddy river, which carries away the moraine rubbish that the glacier deposits at its lower end, in some cases almost as fast as it is formed—perhaps I might rather say, as slowly as it is formed; because if you go day after day, you might see scarcely any difference in the detail of certain moraines, though in time, when favourably placed to be worked upon by water, stones of moderate size, that have been shed from the ice, are carried by the river down the valley. In other cases, however, it happens that from various circumstances moraines are preserved from destruction, and form permanent features in the scenery.

Now I have something special to say about moraine-stones, before I describe the glacial phenomena of our own island. When an immense weight of ice, in some cases thousands of feet in thickness, forming a glacier, passes over solid rocks, by the pressure of the moving mass the rocks in the valley over which the glacier passes become smoothed and polished; not flatly, but in wavy lines, presenting a largely mammillated surface. Furthermore, the stones of the surface moraines frequently fall into crevasses, and the small *débris* and finely-powdered rocks, that more or less cover the surface of a glacier, are also borne into these crevasses by the water that flows upon the surface: the consequence is, that the bottom of a glacier is not simply bare ice, but, between the ice and the rock over which it flows, there are blocks of stone imprisoned, and siliceous and feldspathic *débris* (chiefly worn from the floor itself), which may be likened to emery-powder. The result is that, let the rock be ever so hard, it is in time polished almost as smooth as a sheet of glass; and this polished surface is scratched and grooved by the coarser *débris* that, being imprisoned between the ice and the rocky floor, is pressed along in the direction of the flow of the ice. By degrees, grooves and deep furrows are thus cut in the rock over which the ice passes.

But the stones that are imprisoned between the ice and the rocky floor not only groove that floor itself, but in turn become scratched by the harder asperities of the rock over which they are forced; and thus it happens that many of the stones of moraines are covered with straight 'scratches,' often crossing each other irregularly, so that we are able by this means to tell, independently of the forms of the heaps, whether such and such a mass is a moraine or not.

These indications of the rounding, smoothing, scratching, and grooving of the rocks in lines, coincident with the flow of the glacier, together with old moraine heaps and scratched stones, are so characteristic of all glaciers, that by this means we are able to detect the important fact that the Swiss glaciers were once of far larger dimensions than they are now, and that they have gradually retreated to their present limits. Far below the present ends of the Swiss glaciers, fifty or eighty miles farther down the valleys, we find all the signs I have described, and others besides, frequently as marked as if the glacier had only left the rocks before the existing vegetation began to grow upon their surfaces.

Such being the case in Switzerland, where we have been able to study the action of glaciers in detail, we have next to inquire, is there anything farther to learn in regions where glaciers are on a far greater scale? Those who have read the descriptions of navigators will be aware that in Greenland the average ice-line, as a whole, descends lower and lower as you go northward, till, in the extreme north, the whole country is one universal glacier.

The same universal covering of ice is found in that southern land discovered by Ross, and known as Victoria Land, where the mountains rise, some of them, 10,000, 12,000, and 14,000 feet above the sea, and except here and there, where the cliffs are very steep, the country is covered with a coating of thick ice. In Greenland, where the coast happens to be high and steep, the glaciers break off at the top of the cliffs, and fall in shivered icebergs into the sea; but when valleys fairly open into the sea, then it frequently happens that prodigious glaciers push their way across the land out at sea, and are, in certain cases, twelve or fourteen miles across at their ends.

In the extreme north, the glacier has been described as proceeding out to sea and forming a continuous cliff of ice as far as the eye can reach,—far outside the true rocky coast.

Some of these vast glaciers have been estimated as being, at the very least, 3000 feet in thickness; and great masses of ice breaking away from their ends form icebergs, which, frequently laden with moraine rubbish just like that which covers the glaciers of Switzerland, float out into the Greenland seas, and are carried south by the current along the coast of North America. Some of these bergs are known to float south beyond the parallel of New York, and they have even been seen off the Azores.

Melting by degrees as they come into warmer climates, the stony freight is scattered abroad, here and there, over the bottom of the Atlantic, which thus becomes strewn with erratic blocks, and other *débris*, borne from far northern regions. I shall now apply these remarks in our own island; and having ascertained what are the signs by which a glacier may be known, I shall show that a large part of the British islands has been subjected to glaciation, or the action of ice.

Those who know the mountains of the Highlands of Scotland remember, that though the weather has had a powerful influence upon them, rendering them in places rugged, jagged, and cliffy, yet, notwithstanding this, their general outlines are often remarkably rounded, flowing in greater and smaller mammillated curves; and when you examine the valleys in detail, you also find that in their bottoms and on the sides of the hills the same mammillated structure frequently prevails. These rounded forms are known to those who specially devote themselves to the study of glaciers by the name of 'roches moutonnées,' a name now in general use in England, because it happened that in Switzerland glaciers were first described by authors who wrote in French. Ice-rounded rocks are exceedingly common in many British valleys; and not only so, but the very same kind of grooving and striation, so eminently characteristic of the rocks in the Swiss valleys, also marks those in the Highlands of Scotland, Cumberland, Wales, and other districts in the British islands. Considering all these things, geologists, led, twenty-three years ago, by Agassiz, have by degrees almost universally come to the conclusion that a very large part of our island was, during the 'Glacial Period,' covered, or nearly covered, with a coating of thick ice, and in the same way that the north of Greenland is at present; so that, by the long-continued grinding power of a great glacier, or set of glaciers, nearly universal over the northern half of our country and the high ground of Wales, the whole surface became *moulded by ice;* and the relics of this action still remain strongly impressed on the country, to attest its former power.

It might be unsafe to form this conclusion merely by an examination of such a small tract of country as the British Islands; but when we consider the great Scandinavian chain, and the north of Europe generally, we find that similar phenomena are common over the whole of that area; and in the North American continent, as far south as latitude 38° or 40°, you find, when you remove the soil, or the superficial covering of what is called drift, and get at the solid rock beneath, that almost everywhere it is smoothed and polished, and covered with grooves and striations similar to those of which we have experience among the glaciers of the Alps. I do not speak merely by common report in this matter, for I know it from personal observation, both in the Old World and the New. We know of no power on earth, of a natural kind, which produces these indications except ice; and therefore geologists are justified in attributing them, even on this great continental scale, to its action.

You will presently see that this conclusion is fortified by several other circumstances. Thus in the Alps there is evidence that the present glaciers were once on an immensely larger scale than at present. The proof, as usual, lies in the polished and grooved rocks far removed from the actual glaciers of the present day, and in numerous moraines on a scale so immense, that the largest forming in the Alps in our time are of mere pigmy size when compared with them. The same kind of phenomena occur in the

Himalayah, the Andes, and in almost every northern mountain-chain or cluster, great or small, that has been examined critically; and therefore there can be no doubt that at a late period of the world's history an extremely cold climate prevailed over much larger tracts of the earth's surface than at present, produced by some cause about which there are many vague guesses, but which no one has yet explained.

It was at this period that a great part of what is now the British Islands was covered with ice. I do not say that they were islands at that time; and I think they were not islands, but probably united with the Continent; and the average level of the land may then have been much higher than at present, chiefly by elevation of the whole, and partly because it had not suffered so much degradation. But whether this was so or not, the mountains and much of the Lowlands were covered with a universal coating of ice, probably as thick as that in the north of Greenland in the present day. While this large ice-action was going on, a slow submersion of the land took place; and as it sank, the glaciers, descending to the level of the sea, deposited their moraine rubbish there. Gradually the land sunk more and more, the cold still continuing; till this country, previously united to the Continent of Europe, became a group of icy islands, still covered with snow, and small glaciers, which descended to the sea and broke off in icebergs. These, floating south, deposited their stony freights as they melted. The proof of this is to be found in the detritus which covers so much of Scotland and two-thirds of England, composed of clay and gravels, mixed with stones and great boulders, many of which are scratched, grooved, and striated in the manner of which we have every day experience in the glaciers of Switzerland, Norway, and Greenland. Much of this clay is known as the 'Till' in Scotland; and it was only by very slow degrees that geologists became reconciled to the idea that this Till is nothing but moraine rubbish on a vast scale, formed by those old glaciers that once covered the northern part of our country. In fact, Agassiz, who held these views, and Buckland, who followed him, were something like twenty years before their time; and men sought to explain the phenomena of this universal glaciation by every method but the true one. Mr. Robert Chambers was, I think, the first after Agassiz who asserted that Scotland had been nearly covered by glacier ice; and now the subject is being worked out in all details, thus coming back to the old generalised hypothesis of Agassiz, which is now accepted, or on the very verge of acceptance, by most of the best geologists of Europe and America.

Besides the proofs drawn from the scattered boulders, we know that the country was descending beneath the sea during this glacial epoch for another reason: that here and there in the heart of the moraine matter of the Till there are patches of sand and clay interbedded. The mass indeed is not stratified, because glaciers do not stratify their moraines; but the waves, playing upon them as they were deposited in the sea, here and there arranged portions in a stratified manner; and there occur at rare intervals, in these patches in Scotland, the remains of sea-shells of species such as now occur in the far north. Here, therefore, we have another proof of that arctic climate which, in old times, came so far south.

In Wales we find similar evidence, long since described by myself, of the sea having risen at least 2300 feet upon the sides of the mountains; for Wales, like Scotland, also became a cluster of islands, round which the drift was deposited; and great blocks of stone were scattered abroad, floated out on icebergs that broke from an old system of glaciers, and melted in the neighbouring seas. In this stratified material sea-shells were long ago found in Caernarvonshire, by Mr. Trimmer and myself, from 1000 to 1400 feet above the sea. Erratic blocks of granite, gneiss, feldspathic traps, and of other rocks—some of which came from the Highlands of Scotland, some from the Cumberland mountains, some from the Welsh mountains, and some from the farther region of the great Scandinavian chain—were in the same manner spread over the central counties, and the west and east of England, just like those boulder-beds that are now being formed at the bottom of the Atlantic from the icebergs that float south from the shores of Greenland.

All these marine boulder-drifts are rudely stratified, when viewed on a large scale, and the clays are often interbedded with sand and rounded gravel; but it is remarkable that in most of the great beds of clay that form the larger part of the formation, the stones and boulders that stud the mass are scattered confusedly, and frequently stand on end, like the stones in the moraine-till of Scotland; and this is the case even when associated with sea-shells, which shells prove the true marine nature of what often looks like a mass of heterogeneous rubbish. A great number of these shells occur in a broken state; and from Scotland to Norfolk, on the coast cliffs, they may be found plentifully enough when carefully looked for. Between Berwick and the Humber I have seen them in scores of places, the most plentiful species, as determined for me by Mr. Etheridge, being *Cardium edule, Cyprina islandica, Venus, Dentalium entalis, Tellina, Leda oblonga, Astarte borealis,* and *Saxicana rugosa*. On the west coast, by the Mersey and near Blackpool, they are equally plentiful; and far inland, near Congleton and Macclesfield, I observed the same kind of broken shells 600 feet above the sea, and I understand that Mr. Prestwich observed them in the same region at a height of fully 1200 feet. It is remarkable, also, what a prodigious number of ice-scratched stones occur in this drift, under such conditions that the idea is suggested that they were marked, not by glaciers, but by the agency of coast-ice. As you travel south, you find that the numbers of the kinds of stones increase according to the number of formations you have passed. North of the magnesia limestone district, the fragments to a great extent consist of Silurian old red and carboniferous fragments; then these become mixed with pieces of magnesian limestone; by and by oolitic fragments are added; and in Holderness, in places half the number of stones are of chalk, generally well scratched, and often mingled with broken shells. But it is evident that the low chalk hills of Yorkshire are not of a kind to have given birth to glaciers, and therefore the scratching and distribution of the chalk-stones may have been produced by coast-ice; and the same may be said of other low parts of England, both on the coast and inland, where the drift prevails, the currents which scattered the ice-borne material having on a great scale flowed approximately from north to south.

But England, south of the estuaries of the Severn and the Thames, for the most part, seems all this time to have remained above the waters; for not only is the country in general destitute of drift, but it is only close on the sea near Selsea and Brighton that erratic boulders of granite, &c., have been found, apparently floated from the Channel Islands, or from France.

After a long period of submergence, the country gradually rose again, and the evidence of this I will prove chiefly from what I know of North Wales, although I could easily do the same, by taking you to the Highlands of Scotland.

I shall take the Pass of Llanberis as an example, for there we have all the ordinary proofs of the valley having been filled with glacier-ice. First, then, after the great glacial epoch, the country, to a great extent, sunk below the water, and the drift was deposited, and more or less filled many of the valleys of Wales. When the land had risen again to a considerable height, the glaciers increased in size, although they never reached the immense magnitude which they attained at the earlier portion of the icy epoch. Still they became so large, that such a valley as the Pass of Llanberis was a second time occupied by ice; and the result was, that the glacier ploughed out the drift and loose rubbish that more or less covered the valley. Other cases of the same kind could easily be given, while, on the other hand, in many valleys you find the drift still remaining. By degrees, however, as we approach nearer our own days, for unknown reasons, the climate slowly ameliorated, and the glaciers began to decline, till, becoming less and less, they crept up and up; and here and there, as they died away, they left their terminal and lateral moraines, still in some cases as well defined as moraines in lands where glaciers now exist. Frequently, too, masses of stone, that floated on the surface of the ice, were left perched upon the rounded 'roches moutonnées,' in a manner somewhat puzzling to those who are not geologists; for they lie in such positions, that they clearly cannot have rolled into them from the mountain above, because their resting-places are separated from it by a hollow; and besides, many of them stand in positions so precarious, that if they had rolled down from the mountains, they must, on reaching the points where they lie, have taken a final bound and fallen into the valley below. But when experienced in the geology of glaciers, the eye detects the true cause of these phenomena, and you have no hesitation in coming to the conclusion, that as the glacier declined in size, the errant stones were let down upon the surface of the rocks so quietly and so softly, that there they will lie until an earthquake shakes them down, or until the wasting of the rock on which they rest precipitates them to a lower level. Finally, the climate still ameliorating, the glaciers shrunk farther and farther into the heart of the mountains, until at length, here and there in their very uppermost recesses, you find the remains of tiny moraines marking the last relics of the ice before it disappeared from our country.

Climate of European Drift and Cave Deposits.[*]

We may now ask, with what European deposits does the frozen mud of Siberia, containing the remains of the mammoth in so fresh a state, correspond geologically? Their superficial distribution, and the species of mammalia, as well as the fact that the shells which Middendorf and others observed in them are of living species, seem to connect them chronologically with that paleolithic drift in which flint implements have been detected in England, France, and Italy. The temperature which prevailed in the val-

[*] From Sir C. Lyell's *Principles of Geology*.

leys of the Thames, Somme, and Seine at the era in question was, according to Mr. Prestwich, 20° Fahrenheit colder than now, or such as would now belong to a country from 10° to 15° of latitude more to the north. This estimate is founded on a careful analysis of the land and fresh-water shells which accompany the remains of the mammoth and its associates in the paleolithic alluvium. If we confine our attention to those terrestrial shells which are most commonly buried in the same gravel and sand as the *Elephas primigenius* and *Rhinoceros tichorhinus*, we find them to amount to no less than forty-eight species in the valley of the Thames and its neighbourhood. All but two of these still survive in Britain; these two, *Helix incarnata* and *Helix ruderata*, still inhabit the continent of Europe, and have a great range from north to south. The associated fresh-water shells, more than twenty in number, are also British species; but as they occur, with two or three exceptions, as far north as Finland, their presence is not opposed to the hypothesis of a cold climate, especially as the Limnœcæ are capable of being frozen up, and then reviving again when the river ice melts. At Fisherton near Salisbury, one of the rude flint implements of the earliest stone age was found in drift containing the mammoth and Siberian rhinoceros, together with the Greenland lemming and a *Spermophilus*, a northern form of quadruped allied to the marmot, besides the tiger, hyena, horse, and other extinct and living species; the whole assemblage being confirmatory of the opinion, that the men of the early stone period had often to contend with a climate more severe than that now prevailing in the same parts of Europe. The late Edward Forbes compared the condition of Britain and the neighbouring parts of the Continent during the period next preceding the historical, to the barren grounds of Boreal America, including the Canadas, Labrador, Rupert's Land, and the countries northwards, where the reindeer, musk-ox, wolf, arctic fox, and white bear now live. But we find in some parts of the drift evidence of a conflicting character, such as may suggest the idea of the occasional intercalation of more genial seasons of sufficient duration to allow of the migration* and temporary settlement of species coming from another and more southern province of mammalia, so that their remains were buried in river gravels at the same level as the bones of animals and shells of a more northern climate. If we allow a vast lapse of ages for the accumulation of the drift, we may take for granted that there must have been such changes in climate, owing sometimes to geographical and sometimes to astronomical causes, which will be treated of in the twelfth chapter.

Bones of the hippopotamus, of a species closely allied to that now inhabiting the Nile, are often accompanied in the valley of the Thames and elsewhere by a species of bivalve shell, *Cyrena fluminalis*, now

* The reader will see the cause for this migration as he advances in this work.

living in the Nile and ranging through a great part of Asia as far as Tibet, but quite extinct in the rivers of Europe. Imbedded in the same alluvium with this shell, we find at Grays, in Essex, *Unio litterolis*, a mussel no longer British, but abounding in France in rivers more southern than the Thames. The *Hydrobia marginata* is also a shell sometimes met with in the drift, a species now inhabiting more southern latitudes in Europe. The kind of elephant and rhinoceros accompanying the cyrena at Grays (*E. antiquus* and *R. megarhinus*) are not the same as the mammoth and rhinoceros which occur with their flesh in the ice and frozen mud of Siberia, or in those assemblages of mammalia which have an arctic character in the drift of England, France, and Germany. Some zoologists conjecture that the fossil species of hippopotamus was fitted for a cold climate; but it seems more probable that when the temperature of the river water was congenial to the cyrena above mentioned, it was also suited to the hippopotamus.

Glacial Epoch.—The next step of our retrospect carries us back to what has been called the Glacial Epoch; which, though for the most part anterior to the valley drifts and cave deposits of the paleolithic age above mentioned, was still so closely connected with that period that we cannot easily draw a line of demarcation between them. The dispersion of large angular fragments of rock called erratics over the northern parts of Europe and North America, far from the nearest parent rocks from which they could

have been derived, had long presented a difficult
enigma to geologists before it began to be suspected
that they might have been transported by ice, and at
a period when large parts of the present continents
were submerged beneath the sea.

These blocks are observed to extend in Europe
as far south as latitude 50°, and still farther in
America, or to latitude 40°. It was remarked that
some of them were polished, and striated on one or
more of their sides in a manner strictly analogous to
stones imbedded in the terminal moraines of existing
glaciers in the Alps. In many areas covered with
them, both in America and Europe, the underlying
solid rocks were seen to be marked by similar
scratches and rectilinear furrows, their direction
usually coinciding with the course which the erratics
themselves had taken. As both the smoothing and
striation of the transported fragments, and the surfaces of the rocks *in situ*, were identical in character
with effects recently produced by existing glaciers,
it was at length admitted—but not till after the point
had been controverted for a quarter of a century—
and in direct opposition to the opinion of the earlier
geologists, that the climate which preceded the historical was not only colder as far south as latitude
50° in Europe, and even to 46° in the Alps, but was
marked by such intensity of cold, and such an accumulation of ice, as to be quite without parallel in corresponding latitudes in the present state of the globe,
whether in the northern or southern hemisphere.

Extracted from Page's 'Past and Present Life of the Globe.'

Having passed the middle ages of the earth's history, whose life-species have all, or nearly all, disappeared, we enter upon an epoch whose forms insensibly graduate into those that are now our fellow participators in the great progressional scheme of vitality. In other words, we approach the Cainozoic, or 'recent life period,' which, though but as yesterday compared with the æons of the palæozoic and mesozoic, yet embraces a vast lapse of time, and is necessarily characterised by higher and still advancing forms. We say necessarily characterised; for though science can prove no casual connection between the physical and vital manifestations of the globe, the one set of changes so invariably accompany the other, that we are compelled to regard them as necessary concomitants. And yet, though concomitants in time, they may stand in no relation to each other as cause and effect, but be each an independent phase of that divine creational plan that is still evolving itself around us. We, who but dimly perceive the broken outline of the scheme, can only note the coincidence; those in after ages, of higher intelligence, may succeed in tracing the connection. But whatever that connection, it is now more marked and appreciable, and geologists can associate with almost every fluctuation of condition a change in the accompanying aspects of cainozoic life.

It is now that the more complex forms of an exogenous flora are superadded to the endogens and gymnogens of the mesozoic, and in their more varied forms and higher utilities become not only a fitter ornament for a more varied surface, but a necessary sustenance for a higher and more diversified fauna. The herbs, and shrubs, and trees—the flowers, and fruits, and grains—all that can gladden the senses or satisfy the wants of man and his existing life-comrades, appear with the current epoch; and, by their appearance, again confirm that fitness that ever reigns between the organic and inorganic aspects of creation. In the animal world, the advance is equally apparent; and in orders where no advance appears, a thousand modifications present themselves. Among the protozoans, the calcareous sponges for the most part disappear, their place being taken by those of a horny nature, while the foraminifera are culminating in size and complexity of configuration. The encrinites, with one or two solitary exceptions, have vanished from the waters; and the sea-urchins, so exquisitely preserved in the chalk, are reduced by several of their most beautiful and numerous families. Among the shell-fish, the brachiopods dwindle to a few families; the true bivalves are still on the increase; the gasteropod univalves become dominant in genera and species, while the shell-clad cephalopods that thronged the mesozoic ocean in myriads perish to a solitary genus. The crustaceans become less natatory and more ambulatory in their character; while the insects, so imperfectly preserved in the past, now throng every element—air, earth, and water—in apparently still increasing numbers. The placoids and ganoids, so long the only representatives of ichthyic life, are now on the wane, and the cycloids and ctenoids appear as the prevailing orders. Of the ichthyosaurs and plesiosaurs, that whale-like ruled the ocean, of the megalosaurs and hylæosaurs that tenanted the plain and roamed the forest, and of the pterosaurs that winged the air, not a living trace remains. They are utterly extinguished, and their place is now filled by the crocodiles, lizards, turtles, and serpents of existing nature. The birds, so scantily preserved (though largely indicated) in mesozoic strata, and the mammals represented only by a few insignificant marsupials, now assume the chief importance in the great vital scheme; and last, and highest of all, man himself enters on the stage of being as the crowning form of the current epoch.

To facilitate comparison, it is usual to subdivide the cainozoic into cocene, miocene, pliocene, and pleistocene—that is, into its earliest, less recent, more recent, and most recent life-stages; but enough for our view to treat it in two great sections: the first, when land and sea had a somewhat different distribution from the present; and the second, when they had assumed, within the limits of an appreciable mutation, their existing arrangement. Adopting the familiar phraseology that designates the palæozoic as *primary*, and the mesozoic as *secondary*, we may regard the first section as *tertiary*, and the second as post-tertiary; ever bearing in mind that such distinctions are mere provisional aids to facilitate the comprehension of geological progression. It has been customary, no doubt, for certain geologists, generalising from limited tracts in Europe, to draw a bold line of demarcation between the chalk and tertiary—so bold that not a single species was regarded as passing from the one epoch to the other. This, like many of the early conclusions of the science, is altogether erroneous; and now even in Europe, to say nothing of America, abundant passage-beds have been detected, showing in this instance, as in every other, that abrupt transitions are at the most merely local and limited phenomena. Assuming, then, that the life-forms of the chalk pass insensibly into those of the tertiary, even though in many European areas the cretaceous era was suddenly brought to a close by the violent displacement of the then land and sea, we yet discover a wide difference between the vital aspects of these respective epochs. In the northern hemisphere, the tertiary seas still trend in an easterly and westerly direction, stretching diagonally through what is now Central Europe and Southern Asia, spreading over a large tract of Northern Africa, and covering in North America wide belts of the Southern States. Shut up from the northern currents, that seem to have influenced the chalk seas, and exposed to those which, like the Gulf Stream, partake of a tropical temperature, the climate of the tertiary areas becomes more genial, and is, in the progress of creation, accompanied by a more exuberant flora and fauna. The seas, even in the latitudes of England, teem with southern forms; while the lands, clothed with a vegetation that finds its nearest analogues in the plants of sub-tropical regions, were tenanted by gigantic mammals which, like the elephant, rhinoceros, hippopotamus, tapir, lion, and tiger, now find their head-quarters in the forests and plains of the torrid zone. Extensive lacustrine areas also appear in certain regions, as in Central France, and in their fresh-water forms present, for the first time, a fauna but doubtfully and obscurely represented in former epochs. (With the exception of the estuarine beds of the weald, and the doubtfully estuarine portions of the carboniferous system, we are altogether ignorant of the fresh-water areas of the older epochs. Lake, river, and marsh, must have existed then as now, each peopled by its own distinctive tenantry; but of these forms we have not a single trace, and it is only as we approach the Tertiary epoch that a fresh-water fauna becomes known and appreciable. As we cannot believe in the total obliteration of ancient fresh-water deposits, so we hopefully look forward to important discoveries in this rich and varied section of vitality.)* In fine, we have every type and feature of existing vitality; and the character of the period will, perhaps, be better

* The passages in parentheses in these extracts appear as notes in the original.

indicated by a notice of the forms that have become extinct, than by any description of the whole, which still constitutes, in a great measure, the flora and fauna now flourishing around us.

Separating the early tertiaries—the eocene and the miocene—from the pliocene and pleistocene, when, under the changing condition of sea and land, the climate of the northern hemisphere began to assume a boreal character, we shall shortly glance at the more marked and peculiar aspects of this early period. Wherever we turn, whether to the clays and gravels of the London basin, or to the marls and gypsums of Paris; whether we restrict our review to the south of Europe, or carry it forward to the centre of Asia, we everywhere find in these earlier tertiaries abundant evidence of a warm, temperate, or even subtropical flora. Palm-like leaves and fruits, such as now flourish on the mud-islands of the Ganges (*flabellaria, nipadites, tricarpellites*), leguminous seeds of arboreal growth (*legumenosites*), twigs and leaves of mimosa, laurel, and other plants, whose congeners now find a habitat in southern latitudes, are thickly scattered through these strata. Nor are these the mere twigs and fragments of tropical forests, drifted from afar by gigantic rivers; for associated with the formation are beds of lignite or wood-coal, composed of kindred plants, that must have flourished for centuries on the spots where their remains are now entombed.

And even if the flora gave no certain evidence of the geniality of the climate that then pervaded the parallels of London and Paris, the associated fauna would of itself establish the belief. Gigantic sharks and rays (*lamna, carcharodon, myliobatis*), such as now frequent the Southern Ocean, crocodiles and turtles (*crocodilus, chelone, emys*) in greater specific exuberance than is now known to the zoologist, tapir-like pachyderms (*palæotherium, anoplotherium*), akin to those of the Malayan peninsula and South America, and river hogs (*hyopotamus, chæropotamus*) like those that now wallow in the mire of African rivers, have left their remains in thousands, testifying at once to the warmth of the climate, and to the long continuance of conditions favourable alike to individual growth and to numerical abundance. An exuberance of pachydermatous quadrupeds, foreshadowing in their varied forms the solidungulates and ruminants of a subsequent era, is perhaps the most notable feature of the period; for, though the remains of whale, opossum, mole, bat, and even monkey, have been detected in the earlier tertiaries of Europe, the dominant impress of mammalian life over a larger section of the northern hemisphere was undoubtedly palæotheroid. In the forest, over the plain, and by lake and river-swamp, these curious creatures held supreme sway, simulating every form—sea-cow, tapir, hog, rhinoceros, ass, camel, antelope—and apparently performing every function now assigned to these later and diverse families. During the prevalence of the genial climate that then prevailed, we find not only an unusual flush of terrestrial life, but discover that the fresh-water lakes, the estuaries, and the seas, also teemed with many new and ascending forms.

It is now that we detect in these marls the approximating species of our *lymneæ, paludinæ, plauorbes*, and other fresh-water shells; the terrestrial snails, *helix, pupa, clausilia*, &c., so slenderly represented in former epochs; and in the clays and limestones, increasing congeners of our marine gasteropods (spindle-shells, periwinkles, volutes, and cowries), *fusus, cerithium, natica, voluta, cypræa* — all assuming so recent an aspect, that the conchologist begins to rank them with living species, and to reckon the chronology of strata by the percentage of existing shells. (The terms 'eocene,' 'miocene,' &c. have reference to the percentage of existing shells contained in the different stages of tertiary strata, thus: Pleistocene (most recent), from 90 to 98 of living species; pliocene (more recent), from 60 to 88 of living species. Miocene (less recent), from 20 to 30 of living species. Eocene (dawn of recent), from 1 to 5 of living species.) It is now, too, that the seas swarm with these foraminiferous organisms that attain their meridian in numbers, bulk, and variety, and give rise, by their myriad calcareous cases, to masses of nummulitic limestone, that rival in extent and thickness the limestones of former epochs, or the coral-reefs of the present day. The nummulitic limestone of the Old World, extending for thousands of miles, and many hundred feet in thickness, and the rivalling orbitoidal masses of the New World, are almost entirely the work of these many-celled foraminifera; while thick and extensive strata of *tripoli*, and other siliceous earths, are wholly made up of the moulted shields of the minuter diatoms and infusoriæ.

There is nothing more wonderful in nature than the magnitude of the stony masses elaborated by the lowest of animated creatures—the diatoms, foraminifera, corals, and other microscopic organisms. It seems as if an ordinance, that the nearer a vital approaches the physical, the less the organic is elevated above the inorganic, the more nearly they should resemble each other in the bulk and character of their lithological operations. And yet the elaboration of limestone from marine waters, by the merest vitalised speck of gelatinous matter, is a result that can never be mistaken for that of mechanical or chemical agency. The two things may approximate; they can never be confounded.

Exuberant as the aspect of eocene life may appear, the march of creation is still ever forward. The physical agencies of nature are ever slowly but surely at work. Here the eocene sea is being gradually elevated into shoals and islands; there the fresh-water lake is submerged, and its sediments overlaid by those of marine origin; and here, again, volcanic energy gives birth to new mountain-chains, which interrupt the former currents of the air and ocean, and new external influences begin to prevail. The eocene gradually merges into the miocene, and the miocene into the pliocene. Old forms drop away, new ones begin to take their places; and a flora and fauna indicative of a more temperate climate begin to establish themselves over the latitudes that now constitute the middle regions of Europe, Asia, and North America. And just as we approach existing nature in time, so in space the flora and fauna begin to assume those distributive features that continue to characterise them more or less at the present day. The maples, planes, elms, willows, and other dicotyledonous trees contained in the middle tertiaries of Europe, bear the closest resemblance to those that still adorn her forests; and the elephants, hippopotami, rhinoceroses, bears, lions, and tigers of the Old World find their congeneric predecessors in the tertiary mastodons, mammoths, hippopotami, rhinoceroses, cave-bears, tiger-like machairodi, and camel-like merycotheres of the same hemisphere. In like manner the sloths, ant-eaters, armadilloes, and llamas of South America find their geographical prototypes in the megatheres, mylodons, glyptodons, and macrauchenes, so abundantly fossil in the upper tertiaries of that continent; while even in Australasia the kangaroo is preceded by the gigantic diprotodon, the lace-lizard by the megalanea, and the apteryx and emeu by the palapteryx and dinornis. As the miocene and pliocene epochs advance, the more and more do their fossil forms assimilate to those now peopling the same geographical regions; till the fossil may be said to graduate into the sub-fossil, and the sub-fossil into the species still existing.

In European tertiaries, for instance, we ascend from the eocene or palæotheroid age to the elephantoid or middle tertiary, and from this again to the later age of ruminates—antelopes, deer, and oxen. Connected as Europe has been with the rest of the Old World ever since the earliest tertiary epoch, we might naturally expect to find many species spreading indiscriminately over the other continents of Asia and Africa; but while this undoubtedly occurs, there are camel-like, giraffe-like, and antelope forms—*merycotherium, sivatherium, bramatherium*, and the like—peculiar to the tertiaries of Asia, which point to distinctive geographical distributions of life that obtained so early as the middle and upper tertiary epochs. And these curious forms—the huge camel-like merycothere, and the elephant-antelope sivathere—suggest a peculiarity that runs through many of these tertiary mammals. Thus, while all the mammalian classes, with the exception of man, are less or more represented in the miocene and pliocene strata of the Old World, one feature that stamps the fauna of the period, and renders it noticeable even to unprofessional inquirers is the vast amount of *intermediate* or *inosculating forms*. The horns of a ruminant, with the proboscis of a pachyderm; the prehensile lip and dentition of a pachyderm, with the light proportions of an antelope; the blending of horse, camel, and tapir; the inosculating of camel and giraffe,—these and many other convergin characters, appreciable only by the practised anatomist, are features that distinguish the tertiary mammals as a strange and peculiar fauna. Nor is it alone the more generalised physiological character, but their bulk is also in many instances a marked peculiarity of the period. The gigantic mammoths and mastodons, the huge hippopotami and rhinoceroses, the great cave-bears and cave-lions, the unwieldy megatheres and glyptodons compared with the existing sloths and armadilloes, the macrauchene compared with the llama, the trogontherium with the beaver, the diprotodon with the kangaroo, or the dinornis with the cassowary,—all point to creational phases as the tertiary that have ceased to manifest themselves in the current era. The tusks of the mammoth have been found from twelve to fourteen feet (measuring along their outer curves), those of the existing elephant rarely exceed half that length; the fore limb of the megathere would far outweigh the largest living sloth; the cuirass of the glyptodon would cover more than a score of armadilloes; the full-grown llama would make but a tiny calf to the macrauchene; and the emeu could walk beneath the stride of the extinct dinornis. This preponderance of bulky frame-works, and the number of intermediate forms that serve as connecting links between species now widely separate, are perhaps the most notable features of the tertiary fauna, and are highly suggestive to the physiologist, who, rising above mere description, strives to attain to the higher knowledge of creational method and law.

The diversified latitudes over which tertiary deposits are spread, and the difficulty of assigning a contemporaneity to strata containing few or no species in common, compels the palæontologist often to deal with the details of the respective areas rather than attempt a generalised expression for the whole. Enough for our outline, however, to remark that as sea and land approach their present configuration, the fossil flora and fauna begin in like manner to assume that distinctive impress which now characterises existing nature. As already stated, many of the Old-World forms are unknown in the New; some of those that characterise the tertiaries of India are unknown in the strata of Europe; and only a few, and these during the earlier stages of the period, appear to have anything like a cosmopolitan extension. So also in the earliest or eocene stage the number of existing species are few compared with the extinct: this proportion increases in the middle stages; and as we rise to the uppermost deposits it is often difficult to draw any specific distinction between the fossils they contain and the plants and animals that now flourish on their superficial areas. In the earliest stages the fauna of Europe was characterised by its palæotheres, anoplotheres, xiphodons, river-hogs, alligators, crocodiles, gavials, and turtles; in the middle stages these decline or die out, and deinotheres, mastodons, mammoths, camels, giraffes, cave-bears, lions, and hyænas take their places; while in the upper stages many of these decline, and mammoths, hippotami, rhinoceroses, deer, wild oxen, horse, bears, and tigers, become the dominant features. In like manner, when we turn to Asia, we can trace a similar ascent from the earlier stages, which contain many forms in common with those of Europe, to the middle stages, characterised by their numerous forms of elephant, sivatheres, bramatheres, camels, giraffe, lion, tiger, monkey, crocodiles, and tortoises of enormous magnitude; and from these again to the upper stages, where the mammoth, rhinoceros, urus, horse, ass, and other creatures, lead insensibly to the existing forms of that gigantic continent. In the same way it will be found with Africa, when geology has carried her researches further into that little-known region; and so also it has been found in North America, whose forms bear a wonderful parallelism to those of Europe; while in South America a similar gradation will yet be determined upwards to those Pampean flats, whose pliocene clays and gravels have yielded those wonderful megatheres, mylodons, toxodons, glyptodons, macrauchenes, and other mammals, whose congeneric forms now people, in diminutive scale, the plains and forests and uplands of that exuberant continent. As with the larger continents, so with smaller and more detached areas. The marsupials of Australia have their forerunner in the gigantic diprotodon; the wingless birds of New Zealand were preceded by palapteryx and dinornis; and the still more gigantic œpyornis of Madagascar foreshadows the advent of the ostrich of Africa.

In the elimination of these successive fauna long

ages must have passed away; and during these ages vast physical changes were necessarily effected on the terraqueous relations of the globe. In the northern hemisphere some of the principal mountain-chains—the Alps, Apennines, Carpathians, and Himalayas—had been gradually assuming their ultimate configuration, and the large inland seas that had occupied the central latitudes of Europe, of Northern Africa, Middle India, and Eastern Siberia and China, had been elevated successively into shoals, lake and island, swamp and dry land. Simultaneously with these terraqueous changes the genial temperature that ushered in the eocene period of Europe and America began, stage by stage, to decline, the miocene was marked by more temperate manifestations; and ultimately the pliocene sank into a condition incompatible with the existence of the former flora and fauna. A cold, glacial, and barren period ensued, and under its rigors pliocene life in the northern hemisphere succumbed, and was succeeded by genera and species akin to those that now people the boreal regions.

This ungenial period, generally known in geology as the 'glacial,' 'northern drift,' or 'boulder-clay' epoch, is lithologically characterised by its superficial mounds and masses of drift-sand and gravel, by thick tenacious clays, interspersed indiscriminately with water-worn blocks of all sizes, from mere pebbles to boulders many tons in weight; and by the polished, rounded, and striated surfaces of the subjacent rocks, as if they had been subjected to the long-continued friction of water or ice-borne material, and scratched and furrowed by the passage of the harder and heavier fragments. In Europe, Asia, and North America, down to the forty-fourth or forty-second parallel of latitude, and up to the altitude of two thousand feet, these appearances present themselves, and are inexplicable, unless on the ground of the gradual submergence of the northern hemisphere to that extent, and its subjection to a boreal climate which engendered glaciers on its hills, and drifted, during a brief summer, icebergs laden with rocky *débris* over its waters. The glaciers smoothing, rounding, and grooving the rocks of the higher grounds—the icebergs grinding their way through firth and strait, dropping their burden of mud, sand, and gravel on the sea-bed, or stranding themselves on its shores—complete the necessary arrangements for the production of the geological phenomena of the period. For ages the pliocene lands must have slowly subsided, each step gradually narrowing the boundaries of vegetable and animal life, and driving the surviving species, under the rigours of a deteriorating climate, to higher and higher regions. Race after race would succumb: first the more limited and local, next the more cosmopolitan, and ultimately few of the old flora or fauna would survive, except the more elastic in constitution, and those that had step by step retreated into more southern latitudes.

How long these conditions continued we have no

means of determining in centuries; but, judging from the amount of denudation, the extent and nature of the heterogeneous deposits, as well as from the slow rate of elevation and submergence now going on in known regions, vast periods must have elapsed during the manifestation of this glacial epoch. At length, the downward tendency of these northern latitudes comes to a close; submergence stops, and elevation begins. Slowly, and for long, under a rigorous climate, the lands of Europe, Asia, and North America emerge from the waters. Glaciers still envelop the higher elevations; icebergs, summer after summer, drift over the waters; and the sea, attacking the soft emerging shores, reassorts and redeposits the sands, gravels, and clays of the older glacial epoch. By and by the deposits become fossiliferous, showing that the ocean was tenanted by shell-fish, seals, whales, and other creatures whose habitats are now the icy regions of the arctic circle. Upward, still upward, the land emerges, evincing in its old water-lines and raised beaches the successive steps of its uprise, till ultimately the continents of the northern hemisphere assume, within appreciable limits of current mutation, the configuration and climatology they now present. As the continents emerge and the land-surfaces augment; as new atmospheric and oceanic currents are established, and as the post-tertiary epoch advances, the boreal races retreat farther to the north, some of the old pliocene families again return and spread over European latitudes, and other and newer forms, in the course of creation, begin to appear.   *   *   *   *

It has been argued, no doubt, that in the primeval epochs our earth, in virtue of its internal heat, enjoyed a higher and more uniform climate; and was consequently peopled by a more uniform flora and fauna. This argument, like many others of the earlier geologists, seems altogether without foundation. Granting the existence of a higher internal temperature in pre-vital times, and admitting its influence on the surface, we are still without a shadow of evidence that this interior heat has exercised the least perceptible effect on the climatology of the globe since the deposition of the fossiliferous strata.

(According to Fourier, Hopkins, and other physicists, the internal heat of the globe, which increases at the rate of $1°$ for every sixty feet of depth, does not at present affect the mean superficial temperature more than 1-20th of a degree; and to have had any sensible effect on external climates — say to the extent of $10°$ — this interior heat must have been 200 times its present amount. At that rate the melting-point of lavas would have been reached at a depth of 580 feet, instead of 116,000 feet, as presently estimated, and all the deeper-seated strata must have been fused or rendered crystalline — a condition in which they do not occur even to the depth of 30,000 feet, as many of the cambrian and silurian slates are merely hardened and cleaved, but in no degree metamorphic.) On the contrary, all that we know of the nature and thickness of pre-cambrian sediments, and all that we have learned of the cambrian rocks themselves, preclude the supposition of such an influence beyond the most infinitesimal degree, and compel us to believe that the physical conditions of life have been much the same throughout every period of its existence. It is true that, during the carboniferous period, the oolite, and the earlier tertiaries, certain latitudes of the northern hemisphere appear to have enjoyed a more genial climate than they do now; but the explanation of this we are to seek in the varying distributions of sea and land, the existence of warmer oceanic currents, and other geographical conditions, rather than in any perceptible influence derived from the earth's interior. Nay, as we have warmer and colder regions *in space*, in virtue of the earth's relations to the solar system, so we are inclined to believe we have had warmer and colder periods *in time*, in virtue of some great but unknown cosmical law. The existence of the glacial or boulder epoch over the greater portion of the northern hemisphere —say up to the fortieth or forty-second parallel of latitude—is now admitted on all hands; and as we cannot entertain the idea of cataclysmal irregularity in creation, so we are led to infer the prior occurrence of such glacial periods at determinate times and over determinate areas. The existence of such glacial recurrences has been surmised by several geologists as characterising the periods of the old red sandstone and permian. (Mr. Cumming, in *History of the Isle of Man*, 1848; Mr. Goodwin Austen, *Geological Journal*, 1850; Professor Ramsey, *ibid.* 1855; and the Author, in his *Advanced Text-Book*.) And we may venture to extend them to other systems as probable features of a great cosmical plan.

Thus, looking at the cambrian strata of the northern hemisphere—their angular grits and conglomerates, their extreme paucity of fossil forms, and other features—we are at once reminded of the action of ice and the presence of ungenial conditions. This is followed over the same areas by the more genial and exuberant period of Siluria, which is in turn succeeded by the old red sandstone, whose grits and bouldery conglomerates, as well as paucity of vegetable forms, once more suggest the recurrence of colder influences. Following the old red we have the exuberant flora and fauna of the coal period, again to be succeeded by the scanty life-forms and grits and conglomerates of permia. Again, the trias and oolite of the northern hemisphere are characterised by life-forms that betoken warm and genial conditions, while the chalk that succeeds imbeds water-worn blocks of granite and lignite, which would seem to imply the presence of ice-drift and deposit in seas that were open to boreal influences. Next, the early tertiaries occur over the same areas, marked by plants and animals that indicate a warm and genial climate; and this, in turn, gives place to the well-

CLIMATIC CYCLES SHOWN BY GEOLOGY. 49

known glacial or boulder-drift epoch; once more to be succeeded by the milder influences of the post-tertiary or current era. Throwing these recurrences into diagrammatic form, we appear to have had an alternation of colder and warmer cycles over the northern hemisphere at least; and if such has really

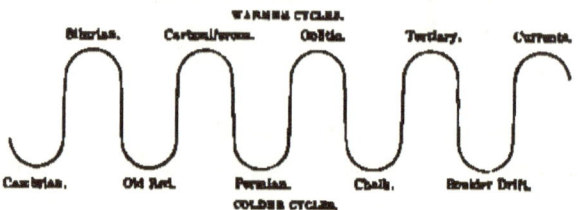

been the case, we must seek the explanation not in revolutions and cataclysms, but in some fixed and continuously operating law. (This idea of colder and warmer cycles, as affecting the northern hemisphere, was indicated some years ago to the Literary and Philosophical Society of St. Andrews. Since then the author has endeavoured to establish the fact, partly by the character and composition of the rocks of the colder periods, and partly by the nature of their fossil contents. Much, however, still remains to be done, and he would earnestly solicit the attention of geologists to the subject; and this, altogether apart from the cosmical causes to which the recurrences may be due. On this latter aspect of the question some discussion took place in the *Athenæum* of 1860, on the suggestion of Colonel James,* of the

* Now Sir Henry James.

Ordnance Survey, that former changes of climate may be due to changes in the inclination of the earth's axis, brought about by alterations in the crust that gradually affect the centre of gravity. Whatever the cause, whether it is to be sought for on or within the globe itself, or in purely astronomical influences, this is not the place to discuss; but, most unmistakably the gradual uprise of land that is now taking place in the Arctic regions, the shifting of volcanic areas in the northern hemisphere since the tertiary period, and the approach and departure of the boulder epoch over the same latitudes, all point to the operation of some determinate law of secular succession. May it not be, that in the periodicity of this law we may yet discover the key to the expression of geological chronology in years and centuries?) Whether the phenomena may depend on causes operating on and within the globe itself, so as to change the axis of rotation, or whether it may not more likely depend on forces purely astronomical, are questions that lie beyond the scope of the present sketch. Enough for our purpose to have indicated the probability of such recurrences, and derive therefrom the conclusion, that the conditions of life have been very much the same through all geological periods—successively varying in different areas, but never presenting, any more than they do now, an universal similitude, and that least of all through the influence of the earth's internal temperature.

TIME GEOLOGICAL.

From Page's 'Past and Present Life of the Globe.'

Whatever may have been the creational development of plants and animals—whenever the advent or whatever the first condition of the human race— the groups and systems of geology afford irrefragable evidence of the lapse of vast epochs of TIME. The idea of immense duration is at once suggested by an examination of the stratified rocks. The innumerable alternations of their shales, limestones, sandstones, and conglomerates—their vast thickness—their repeated laminations—the alternation of marine and fresh-water beds—their upheaval into dry land, and subsequent submergence again and again—the various races that have lived and grown and been entombed in them, system after system—all this, and much more that will readily suggest itself to the reflecting mind, must clench, beyond cavil, the conviction of the unconceivable duration of geological time. In all our reasonings, then, we must never lose sight of the element TIME. With unlimited duration at command, we have a power equal to the mightiest results; and forces, which in themselves appear puny and feeble, become giants when backed by that spirit of unrest whose eye never closes, whose wing never wearies, and whose foot never tires. The hardest rock is hollowed by the ceaseless water-drop; the Nilotic plain has been borne, particle by particle, from the mountains of Abyssinia; and the massive coral reef of a thousand leagues owes its origin to an animated speck all but invisible to the unassisted eye. The problems of geology, like the problems of mechanics, are thus dependent for their solution on the conjoint elements of *force* and *time*. Where the exertion of force is great, the time for performance may be short; but where the force is small, the time must be proportionally prolonged. The two elements are ever in inverse ratio; and thus agents in themselves comparatively insignificant may during the lapse of ages accomplish most important results. It is generally, therefore, to the cumulative effects of this inexhaustible resource of TIME that the developists make their last appeal, contending that the progress of transmutation is so gradual as not to be appreciable with the five or six thousand years of man's observation. Admitting the plausibility of the argument in existing nature, the geologist appeals to the fossil world for evidence of these insensible gradations, and he finds stratum after stratum containing the same unchanged species; and then in the next stratum at once and decidedly the remains of a species altogether new. This, if anything we can intelligibly define, is not genetic gradation, but pre-appointed creation; and, as time is merely passive unless the law of specific progress obeys some active and controlling power, an eternity of time will never affect it.

In the present state of our knowledge any attempt to calculate geological knowledge by years is altogether futile; we can only indicate its vastness by the use of indefinite terms, as 'eras' and 'epochs' and 'cycles.' It is customary, however, to speak of *pre-geological, geological,* and *historical* time — meaning by pre-geological all that extends backwards before the deposition of the fossiliferous rocks; by geological, all that is embraced between the earliest fossiliferous deposits and human history; and by historical, all to which a determined chronological value can be assigned. With regard to the first, it is an abysm which the human intellect, even in its boldest flights, shrinks from exploring; as to the last, important as it may seem to man, creationally it is but a thing of yesterday; while to time geological we turn as entering into every problem of our science, and investing their consideration with strange and deeper interest. The amount of this time we have as yet no means of estimating — no power to give it expression in years and centuries. Many ingenious calculations have, no doubt, been made to approximate the dates of certain geological events; but these, it must be confessed, are more amusing than instructive. For example, so many lines of mud are annually laid down by the inundation of the Nile; fragments of pottery have been found at the depth of thirty feet: how many years have elapsed since the pottery was first embedded? Again, the ledges of Niagara are wasting at the rate of so many feet per century: how many years must the river have taken to cut its way back from Queenstown to the present Falls? Again, lavas and melted basalts cool, according to the size of the mass, at the rate of so many degrees in a given time: how many millions of years must have elapsed (supposing an original igneous condition of the earth) before its crust had attained a state of solidity? or farther, before its surface had cooled down to the present mean temperature? For these and similar computations it will at once be perceived that we want the necessary uniformity of factor; and until we can bring elements of calculation as exact as those of astronomy to bear on geological chronology, it will be better to regard our 'eras' and 'epochs' and 'cycles' as so many terms indefinite in their duration, but sufficient for the magnitude of the operations embraced within their limits. But even on this point of expressible time the earnest geologist is not without hope and encouragement. He rests confident (confident as in the existence of his own being) that the whole history of geological phenomena—the shifting of volcanic energy from centre to centre, the elevation and depression of certain areas of the earth's crust, the interchanges of sea and land thereby occasioned, the recurrence of colder and warmer climates over determinable latitudes, the necessary re-arrangements of life attending these changes, and the like—is but a chronological exposition of the influence of natural law; and, that as law is as obedient to *times* as to *modes*, the periodicity of these occurrences will one day or other be determined. This done, its expression in years and centuries is a simple task; but though accomplished tomorrow, and expressed in figures like the distances of the astronomer, the mind would altogether fail to grasp the conception of its immensity.

Extracted from the 'Gallery of Nature.' (Milner.)

Scattered through the gravel, clay, and sand of the drift, and frequently occurring in an independent manner, there are masses of rock denominated erratic, from being found at a distance from the place of their origin. Owing to alluvial agency having removed the light accompanying débris from these blocks, they appear insulated upon the surface, sometimes forming rocking-stones, being so poised that a small force will make them oscillate. The size of some of these erratic blocks is enormous. That out of which the pedestal of the statue of Peter the Great at Petersburg was hewn weighed 1500 tons, and was an insulated drifted mass of granite, that lay on a marshy plain near the city, from whence it was removed on rollers and cannon-balls while the ground of the marsh was hard frozen. The Needle-mountain in Dauphiny, described as erratic, is 1000 paces in circumference at the top, and 2000 at the base. Near Neufchâtel a mass of granite occurs forty feet high, fifty long, and twenty broad, estimated to weigh 3,800,000 pounds. On the Jura limestone mountains, at an elevation of 2000 feet above the Lake of Geneva, there are granitic blocks the solid contents of which amount to 1200 cubic feet upon the Salere, to 2250 on the Côteau de Boissy; and the block called Pierre à Martin is mentioned by Mr. Greenough as containing even 10,296 cubic feet. The rock in Horeb, from which, according to monkish tradition, the leader of Israel miraculously drew water, is a mass of sienitic granite six yards square, containing 5823 cubic feet, lying insulated upon a plain near Mount Sinai. Professor Hitchcock remarks of the United States: 'In this country boulders occur of equal dimensions. Thus on Cape Ann and its vicinity I have not unfrequently met with blocks of sienite not less than thirty feet in diameter; and in the south-east of Bradford county I noticed one thirty feet square, which contains 27,000 cubic feet, and weighs not less than 2310 tons. In the west part of Sandwich, on Cape Cod, I have seen many boulders of granitic gneiss twenty feet in diameter, which contain 8000 cubic feet, and weigh as much as 680 tons. Two greywacke boulders of the same size lie a few rods distant from the meeting-house in Norton, in Dr. Bates's garden. A granite boulder of equal dimensions lies about half a mile south-east of the meeting-house in Warwick; and one of similar dimensions lies on the western slope of the Hosac mountain, at least 1000 feet above the valley over which it must have been transported. One of granite lies at the foot of the cliffs at Gay Head, on Martha's vineyard, which is ninety feet in circumference, and

EVIDENCE OF DRIFT.

weighs 1447 tons. At Fall River there is a boulder of conglomerate, which originally weighed 5400 tons.'

But the instances are endless that might be given of similar large and insulated masses now lying at a remote distance from the parent rocks from which they have been abstracted. Wandering in the southern extremity of America, Mr. Darwin speaks of his course being obstructed by well-rounded pebbles of porphyry, mingled with immense angular fragments of basalt, and of primary rocks, nearly seventy miles apart from the nearest analogous mountain; while in Europe, the vast plains of Prussia, Poland, and Russia, comprehending a zone of country near two thousand miles long, and from four hundred to eight hundred miles wide, are strewed with loose detritus and crystalline blocks of colossal size, all of which have been transported from the mountain masses of Scandinavia. The facts ascertained respecting the dispersion of drift and erratic blocks comprise the following phenomena:

1. Evidence appears of the drift, using the term comprehensively, having been carried outward from the summits and axes of particular mountains and spread over the plains and valleys in their neighbourhood. The best example of this occurs in the Alps, where boulders have been usually carried down the valleys, and exist in the greatest abundance opposite the lower openings of those valleys. Similar instances have been pointed out among the mountains of Scotland and those of the north of England.

The plain of Metidja on the south of Algiers, is described by Rozet as covered in its northern parts by boulders derived from a long chain of hills running along its northern border, while its southern region is strewed with blocks from the Atlas chain, which stretches along its southern limits.

2. Viewed upon a larger scale the drift affords evidence of having been dispersed in a general southerly direction over all the northern hemisphere, often to a great distance, variously modified, however, by local obstructions. Dr. Buckland mentions pebbles and blocks of granite and sienite of a very peculiar character drifted from the Criffle mountain in Galloway, across the Solway Firth, and scattered over the plain of Carlisle; while pebbles and large blocks of another kind of granite have been drifted in still greater numbers from Ravenglass, on the west of Cumberland, over the plains of Lancashire, Cheshire, and Staffordshire, upon which they lie in masses of several tons weight. Professor Sedgwick states that the blocks of shap granite near Penrith, which cannot be confounded with other rocks in the north of England, are not only drifted over the hills near Appleby, but have been scattered over the plains of the new red sandstone, rolled over the great central chain of England into the plains of Yorkshire, imbedded in the transported detritus of the Tees, and even carried to the eastern coast, a direction from the parent bed south by east. Another current of drift found in the east of England, consisting of varieties

of primitive and transition rocks, which do not occur in this country, the origin of which must be referred to Norway, has pursued a course south by west. On the Continent the evidence of a southerly direction is very apparent, the blocks and pebbles that are strewed over the plains of North Germany, Poland, and Russia, having their parent rocks in Sweden, Lapland, and Finland. Across the Atlantic the boulders spread over the southern part of Nova Scotia have been derived from the ledges in the northern part of the province; and throughout Maine and Massachusetts the direction taken by the drift, as shown by a multitude of examples, is from north to south, varying a few points towards south-east or south-west. In many parts of the States south of the western lakes the surface is bestrewed with an immense number of fragments of primitive rocks, significantly called 'lost rock,' which may be traced to the north side of the lakes in Upper Canada. The great valley of the Missouri and Mississippi, from the Yellow Stone river almost to the Gulf of Mexico, is represented by Mr. Catlin as covered with vast quantities of blocks of primary rocks, which have been drifted thither from the north-west. He describes five remarkable blocks of granite from fifteen to twenty-five feet in diameter, on the Côteau des Prairies to the west of Lake Superior, several hundred miles from the nearest rocks of that kind in place which lies northward.

3. Mr. Darwin affirms that in the intertropical

parts of America, Africa, and Asia, erratic blocks have never been observed, nor at the Cape of Good Hope, nor in Australia. Though this complete limitation of them to the colder regions of the globe may be deemed an unsettled point, it is certain that such phenomena are far less frequent and extensive in equatorial countries. Beyond 41° of south latitude in South America, the appearance of drift is common upon the vast plains of Patagonia and Chili.

4. Rocks *in situ*, or in place, very frequently exhibit a marking with fine linear striæ or scratches, larger grooves, and even furrows, with a smooth and polished aspect of the surface, like that which the rapid passage of heavy masses over them would produce. Transatlantic facts of this description have been carefully collected by Professors Hitchcock and Rogers, from whom a few particulars may be introduced. In various parts of Massachusetts the striæ are very obvious and distinct, frequent on the hard sienitic rocks, though often these are merely smoothed and sometimes polished. They are visible on the gneiss at the top of the Wachusett mountain, the highest in the eastern part of the State, two thousand feet above the ocean. The precipitous hills and the lower grounds of the Connecticut river valley are covered with them, their direction being north and south, inclining a few points east of south and west of north. Mount Everett, 2600 feet above the sea-level, has been worn over its whole surface, and the striæ are very visible in many places, though so long exposed naked

to atmospheric and decomposing agencies. Mount Monadnoc, 3250 feet high, little else than a naked mass of mica slate of peculiar character, and almost destitute of stratification, has been from top to bottom scarified and scratched, the striæ running about north-west and south-east. Similar markings appear on the summits of all the Appallachian chain in Pennsylvania, which observe the same direction, and particularly around the Wyoming valley the tops of the mountains are covered with parallel striæ. It is impossible, says one of the authorities named, to stand upon these lofty and precipitous ridges and witness this phenomenon without being struck with the great power and extent of the agency that has thus left its traces upon some of the most elevated spots in New England. The same appearances have been noticed in North Wales, on the Corstorphine hills near Edinburgh, and other parts of Scotland, in Sweden and North Russia. In some instances the tops of ledges of rock appear to have been crushed and bent obliquely by the pressure upon them of an enormous load in movement. The occurrence of drift and of erratic blocks constitutes one of the most complicated and difficult problems of geology: nor has any theory yet been proposed offering a solution of it which is perfectly satisfactory. Some reasons have already been assigned for rejecting the long-prevalent opinion that these phenomena resulted from the deluge of Noah, namely, the presence of animals of extinct species and some of extinct genera in the drift, with the absence of the remains of human beings and of all vestiges of the arts and operations of man; and in addition it may be remarked that if that event was local, confined to a part of south-western Asia, as we have previously argued, it is further inadmissible as an explanation of effects certainly common to the whole northern hemisphere and to the southern extremities of America. But the single consideration, that the vast masses of gravel and blocks strewed over such extensive areas do not belong to one violent and transitory period, but have been aggregated at different epochs by causes sometimes operating feebly and slowly, and at other times violently and powerfully—the evidence of which is irresistible to those who have studied the subject—is amply sufficient by itself to disprove the idea of these accumulations being the consequence of the transient flood recorded in the sacred history.

Extracted from a Paper on 'The Erratic Phenomena about Lake Superior.' By LOUIS AGASSIZ.
(Boston, 1850.)

So much has been said and written within the last fifteen years upon the dispersion of erratic boulders and drift both in Europe and America, that I should not venture to introduce this subject again if I were not conscious of having essential additions to present to those interested in the investigation of these subjects.

It will be remarked by all who have followed the discussions respecting the transportation of loose materials over great distances from the spot where they occurred primitively, that the most minute and most careful investigations have been made by those geologists who have attempted to establish a new theory of their transportation by the agency of ice.

The part of those who claim currents as the cause of this transportation has been more generally negative, inasmuch as, satisfied with their views, they have generally been contented simply to deny the new theory and its consequences, rather than investigate anew the field upon which they had founded their opinions. Without being taxed with partiality, I may, at the outset, insist upon this difference in the part taken by the two contending parties; for, since the publication of Sefstroem's paper upon the drift of Sweden, in which very valuable information is given respecting the phenomena observed in that peninsula, and the additional data furnished by De Verneuil and Murchison upon the same country and the plains of Russia, the classical ground for erratic phenomena has been left almost untouched by all except the advocates of the glacial theory. I need only refer to the investigations of M. de Charpentier, Escher, Von Derlinth, and Studer, and more particularly to those extensive and most minute researches of Professor Guyot in Switzerland, without speaking of my own, and some contributions from visitors— as the Martins, James Forbes, and others—to justify my assertion that no important fact respecting the loose materials spread all over Switzerland has been added by the advocates of currents since the days of Saussure, De Luc, Escher, and Von Buch; whilst Professor Guyot has most conclusively shown that the different erratic basins in Switzerland are not only distinct from each other, as was already known before, but that in each the loose materials are arranged in well-determined regular order, showing precise relations to the centres of distribution from which these materials originated; an arrangement which agrees in every particular with the arrangement of loose fragments upon the surface of any glacier, but which no cause acting convulsively could have produced.

The results of these investigations are plainly that the boulders found at a distance from the central Alps originated from their higher summits and valleys, and were carried down at different successive periods in a regular manner, forming uninterrupted walls and ridges, which can be traced from their starting-point to their extreme peripheric distribution.

I have myself shown that there are such centres of distribution in Scotland, England, and Ireland; and these facts have been since traced in detail in various parts of the British isles by Dr. Buckland, Sir Charles Lyell, Mr. Darwin, Mr. M'Lachlan, and Professor James D. Forbes, pointing clearly to the main mountain groups as to so many distinct centres of dispersion of these loose materials. Similar phenomena have been shown in the Pyrenees, in the Black Forest, and in the Vosges; showing beyond question that, whatever might have been the cause of the dispersion of erratic boulders, there are several separate centres of their distribution to be distinguished in Europe. But there is another question connected with this local distribution of boulders which requires particular investigation, the confusion of which with the former has no doubt greatly contributed to retard our real progress in understanding the general question of the distribution of erratics. It is well known that Northern Europe is strewed with boulders, extending over European Russia, Poland, Northern Germany, Holland, and Belgium. The origin of these boulders is far north, in Norway, Sweden, Lapland, and Liefland; but they are now diffused over the extensive plains west of the Ural Mountains. Their arrangement, however, is such that they cannot be referred to one single point of origin, but only in a general way to the tracts of land which rise above the level of the sea in the Arctic regions. Whether these boulders were transported by the same agency as those arising from distinct centres on the main continent of Europe, has been the chief point of discussion. For my own part, I have indeed no doubt that the extreme consequences to which we are naturally carried by admitting that ice was also the agent in transporting the northern erratics to their present positions, has been the chief objection to the view that the Alpine boulders have been distributed by glaciers. It seemed easier to account for the distribution of the northern erratics by currents; and this view appearing satisfactory to those who supported it, they at once went farther, and opposed the glacial theory even in those districts where the glaciers seemed to give a more natural and more satisfactory explanation of the phenomena. To embrace the whole question, it should be ascertained, *first*, whether the northern erratics were transported at the same time as the local Alpine boulders, and if not, which of the phenomena preceded the other; and, again, if the same cause acted in both cases, or if one of the causes can be applied to one series of these phenomena, and the other cause to the other series. An investigation of the erratic phenomena in North America seems to me likely to settle this question, as the northern erratics occur here in an undisturbed continuation over tracts of land far more extensive than those in which they have been observed in Europe. For my own part, I have already traced them from the eastern shores of Nova Scotia through New England and the north-western States of North America and the Canadas, as far as the western extremity of Lake Superior, a region embracing about 30° of longitude. Here, as in Northern Europe, the boulders evidently originated farther north than their present location, and have been moved universally in a main direction from north to south.

From data which are, however, rather incomplete,

it can be further admitted that similar phenomena occur further west, across the whole continent, everywhere presenting the same relations. This is to say, everywhere pointing to north as to the region of the boulders, which generally disappear about latitude 38°.

Without entering at present into a full discussion of any theoretical views of the subject, it is plain that any theory, to be satisfactory, should embrace both the extensive northern phenomena in Europe and North America, and settle the relation of these phenomena to the well-authenticated local phenomena of Central Europe. Whether America itself has its special local circumscribed centres of distribution or not, remains to be seen. It seems, however, from a few facts observed in the White Mountains, that this chain, as well as the mountains of north-eastern New York, had not been exclusively, and for the whole duration of the transportation of these materials, under the influence of the cause which has distributed the erratics through such wide space over the continent of North America. But whether this be the case or not—and I trust local investigations will soon settle the question—I maintain that the cause which has transported these boulders in the American continent must have acted simultaneously over the whole ground which these boulders cover, as they present throughout the continent an uninterrupted sheet of loose materials, of the same general nature, and evidently dispersed at the same time.

Moreover, there is no ground at present to doubt

the simultaneous dispersion of the erratics over the Northern Europe and Northern America. So that the cause which transported them, whatever it may be, must have acted simultaneously over the whole tract of land west of the Ural Mountains and east of the Rocky Mountains, without assuming anything respecting Northern Asia, which has not yet been studied in this respect; that is to say, at the same time over a space embracing 200° of longitude. Again, the action of this cause must have been such —and I insist strongly upon this point as a fundamental one—the momentum with which it acted must have been such that, after being set in motion in the north with a power sufficient to carry the large boulders which are found everywhere over this vast extent of land, it vanished, or was stopped after reaching the 35° of northern latitude.*

Now it is my deliberate opinion that natural philosophy and mathematics may settle the question whether a body of water of sufficient extent to produce such phenomena can be set in motion with sufficient velocity to move all these boulders, and nevertheless stop before having swept over the whole surface of the globe. Hydrographers are familiar with the actions of currents, with their speed, and with the power with which they can act. They know also how they are distributed over our globe. And if we institute a comparison, it will be seen

* As the reader advances in this work he will be able to judge how ably this evidence has been collected by Mr. Agassiz.

that there is nowhere a current running from the poles towards the lower latitude, either in the northern or southern hemisphere, covering a space equal to one-tenth of the currents which should have existed to carry the erratics into their present position. The widest current is west of the Pacific, which runs parallel to the equator, across the whole extent of that sea from east to west, and the greatest width of which is scarcely 50°. This current, as a matter of course, establishes a regular rotation between the waters flowing from the polar regions towards lower latitudes.

The Gulf Stream, on the contrary, runs from west to east, and dies out towards Europe and Africa, and is compensated by the currents from Baffin's Bay and Spitzbergen, emptying into the Atlantic, while the current of the Pacific, moving towards Asia, and carrying floods of water in that direction, is maintained chiefly by antarctic currents, and those which follow the western shore of America from Behring's Straits. Wherever they are limited by continents, we see that the water of these currents, even when they extend over hundreds of degrees of latitude, as the Gulf Stream does in its whole course, are deflected where they cannot follow a straight course.

Now, without appealing with more detail to the mechanical conditions involved in this inquiry, I ask every unprejudiced mind acquainted with the distribution of the northern boulders, whether there was any geographical limitation to the supposed northern current to cause it to leave the northern erratics of Europe in such a regular order, with a constant bearing from north to south, and to form on its southern termination a wide regular zone from Asia to the western shores of Europe, north of the 50th degree of latitude, before it had reached the great barrier of the Alps? I ask whether there was such a barrier in the unlimited plains which stretch from the Arctic Sea uninterrupted over the whole northern continent of America, as far down as the Gulf of Mexico? I ask, again, why the erratics are circumscribed within the northern limits in the temperate zone, if their transportation is owing to the action of the water-current? Does not, on the contrary, this most surprising limit within the arctic and northern temperate zones, and in the same manner within the antarctic and southern temperate zones, distinctly show that the cause of transportation is connected with the temperature or climate of the countries over which the phenomena were produced? If it were otherwise, why are there no systems of erratics with an east and west bearing, or in the main direction of the most extensive currents flowing at present over the surface of our globe? It is a matter of fact—of undeniable fact, for which the theory has to account—that in the two hemispheres the erratics have direct reference to the polar regions, and are circumscribed within the arctics and the colder parts of the temperate zone. This fact is as plain as the other fact, that the local distribution of boulders has reference to high mountain ranges, to groups of land raised above the level of the sea into heights the temperature of which is lower than the surrounding plains. And what is still more astonishing, the extent of the local boulders, from their centre of distribution, reaches levels the mean annual temperature of which corresponds in a surprising manner with the mean annual temperature of the southern limit of the northern erratics.

We have therefore in this agreement a strong evidence in favour of the view that both the phenomena of local mountain erratics in Europe, and of northern erratics in Europe and America, have probably been produced by the same cause.

The chief difficulty is in conceiving the possibility of the formation of a sheet of ice sufficiently large to carry the northern erratics into their present limits of distribution. But this difficulty is greatly removed when we can trace, as in the Alps, the progress of the boulders under the same aspect from the glaciers now existing down into regions where they no longer exist, but where the boulders, and other phenomena attending their transportation, show distinctly that they once existed. Without extending farther this argumentation, I would call the attention of the unprejudiced observer to the fact that those who advocate currents as the cause of the transportation of erratics have up to this day failed to show in a single instance that currents can produce all the different phenomena connected with the transportation of the boulders which are observed everywhere in the Alps, and which are still daily produced there by the small glaciers yet in existence. Never do we find that water leaves the boulders which it carries along in regular walls of mixed materials: nor do currents anywhere produce upon the hard rocks *in situ* the peculiar grooves and scratches which we see everywhere under the glacier and within the limits of their ordinary oscillations.

Water may polish the rocks, but it nowhere leaves straight scratches upon their surface: it may furrow them, but these furrows are sinuous, acting more powerfully upon the soft parts of the rocks or fissures already existing, while glaciers smooth and level uniformly the hardest parts equally with the softest, and, like a hard file, rub to uniform continuous surfaces the rocks upon which they move. But now let us return to our special subject—the erratics of North America.

The phenomena of drift are more complicated about Lake Superior than I have seen them anywhere else; for, besides the general phenomena which occur everywhere, there are some peculiarities noticed which are to be ascribed to the lake as such, and which we do not find in places where no large sheet of water has been brought into contact with the erratic phenomena. * * * Now that these phenomena have been observed extensively, we may derive also some instruction from the limits of their geographical extent.

Let us see where these polished, scratched, and furrowed rocks have been observed.

In the first place, they occur everywhere in the north within certain limits of the arctics, and through the colder parts of the temperate zone. They occur also in the southern hemisphere, within parallel limits; but in the plains of the tropics, and even in the warmer parts of the temperate zone, we find no trace of these phenomena, and nevertheless the action of currents could not be less there, and could not at any time have been less there than in the colder climates. It is true, similar phenomena occur in Central Europe, and have been noticed in Central Asia, and even in the Andes of South America; but these always in higher regions at definite levels above the surface of the sea, everywhere indicating a connection between their extent and the colder temperature of the places over which they are traced.

More recently a step towards the views I entertain of this subject has been made by those geologists who would ascribe them to the agency of icebergs.* Here, as in my glacial theory, ice is made the agent: floating ice is supposed to have ground and polished the surfaces of the rocks, while I consider them to have been acted upon by terrestrial glaciers. To settle this difference we have a test which is as irresistible as the other arguments already introduced.

* See also a remark copied in the *Times*, October 1872, from Professor Agassiz relative to Ice action in South America.

Phenomena analogous to those produced by icebergs would only be seen along the seashores; and if the theory of drifted icebergs were correct, we should have, all over those continents where erratic phenomena occur, indications of retreating shores as far as the erratic phenomena are found. But there is no such thing to be observed over the whole extent of the North-American continent, nor over northern Europe and Asia, as far as the northern erratics extend. From the arctics to the southernmost limits of the erratic distribution we find nowhere the indications of the action of the sea as directly connected with the production of the erratic phenomena. And wherever the marine deposits rest upon the polished surfaces of ground and scratched rocks, they can be shown to be deposits formed since the grooving and polishing of the rocks, in consequence of the subsidence of those tracts of land upon which such deposits occur.

Again, if we take for a moment into consideration the immense extent of land covered by erratic phenomena, and view them as produced by drifted icebergs, we must acknowledge that the icebergs of the *present period* at least, are insufficient to account for them, as they are limited to a narrower zone. And to bring icebergs in any way within the extent which would answer for the extent of the distribution of erratics, we must assume that the northern ice-fields from which these icebergs could be detached and float southwards were much larger at the time they produced such extensive phenomena than they are now. That is to say, we must assume an ice period; and if we look into the circumstances, we shall find that this ice period, to answer to the phenomena, should be nothing less than an extensive cap of ice upon both poles. This is the very theory which I advocate; and unless the advocates of an iceberg theory go to that length in their premises, I venture to say, without fear of contradiction, that they will find the source of the icebergs fall short of the requisite conditions which they must assume, upon due consideration, to account for the whole phenomena as they have really been observed.

The following is from Darwin's 'Origin of Species.'

But we must return to our more immediate subject, the glacial period. I am convinced that Forbes's view may be largely extended. In Europe we have the plainest evidence of the cold period, from the western shores of Britain to the Ural range, and southwards to the Pyrenees. We may infer from the frozen mammals and nature of the mountain vegetation that Siberia was similarly affected. Along the Himalaya, at points nine hundred miles apart, glaciers have left the marks of their former low descent; and in Sikkim Dr. Hooker saw maize growing on gigantic ancient moraines. South of the equator we have some direct evidence of former glacial action in New Zealand, and the same plants found on widely

separated mountains in that island tell the same
story. If one account which has been published can
be trusted, we have direct evidence of glacial action in
the south-eastern corner of Australia.

Looking to America in the northern half, ice-
borne fragments of rock have been observed on the
eastern side as far south as lat. 36°, 37°; and on the
shores of the Pacific, where the climate is now so
different, as far south as lat. 46°, erratic boulders
have also been noticed on the Rocky Mountains. In
the Cordillera of equatorial South America glaciers
once extended far below their present level. In Central Chili I examined a vast mound of detritus, crossing the Portillo valley, which I now fully believe to
have been due to ice-action. Along this whole space
of the Cordillera true glaciers do not now exist even
at much more considerable heights. Farther south
on both sides of the continent from lat. 41° to the
southernmost extremity, we have the clearest evidence of former glacial action in huge boulders transported far from their parent source.

*From the article 'Geology' in the Encyclopædia
Britannica.*

A change now took place of a kind different from
any we have met with, unless Professor Ramsey's
ideas as to the glacial origin of the permian and other
old conglomerates be well founded. Simultaneously
with a gradual but eventually a great and wide-spread

depression of land, amounting in many places to 2000 or 2500 feet, there was a refrigeration of the climate of our latitudes, so that the glens of our present mountains were encumbered with glaciers, even where their valleys were penetrated by the sea, and our lowlands were entirely submerged. By the action of the glaciers the rocks were scarred and rounded, polished and grooved, and masses of rock carried down and heaped into moraines, while great blocks of rock were transported in fragments of those glaciers which dipped into the sea and formed icebergs, being often carried far over the shallow seas, and dropped many miles from their parent sites, sometimes resting on hill-tops, which were then banks and shallows of the sea, and so arresting the icebergs in their course; alternations of elevation and depression doubtless took place, and the ordinary action of the breakers along the beach was aided by the quantity of detritus poured into it by the glaciers, and modified by that of the shore-ice which formed along it in the winter seasons.

These glacial deposits are not confined to the British Islands, but extend over all the North of Europe and North America, down to a curved boundary which in Europe, according to Sir R. Murchison, only stretches so far south as 50° lat. in one part of its course, namely, near Cracow. Great blocks of Swedish or Norwegian rocks as large as cottages lie scattered over the plains of North Germany.

• • • • •

Much valuable and detailed information relative to the drift in Northern Russia will be found in Sir R. Murchison's work, 'Russia and the Ural;' whilst in the writings of nearly every geologist reference is made to the glacial epoch. Enough, however, has we believe been given in the preceding pages to place before the reader the main facts connected with this interesting and late change on the earth's surface.

CHAPTER III.

CONSIDERATION OF THE CONJECTURED CAUSES SUPPOSED TO ACCOUNT FOR THE GLACIAL EPOCH.

We may now gather from the evidence which has been brought forward the most important facts revealed by the study of geology, and investigate these in a compact form.

It appears that just previous to the present existing conditions there was a singular climate on earth. The effect of this climate was such as not only to produce ice and immense icebergs in latitudes within 45° of the Equator, but also to cause these icebergs to travel in various directions. This climate was neither partial nor local; but extended from the west of North America to the east of Russia, from the plains of Australia to the land above Cape Horn. The equatorial regions of the earth alone seem to have escaped this climate.

The effects of the glacial period are also shown by the remains of glaciers in districts such as Wales and Scotland, where glaciers do not now exist. Also it appears that in various mountainous districts in Europe there is evidence of the glaciers having extended far beyond their present limits, *both upwards and downwards*. It is also considered most probable that our large inland lakes are in a great measure due to the melting of masses of snow and ice formed during the cold of the boulder period.

There is evidence on the rocks of the scratching and grinding produced by large and heavy masses of ice passing over their surface, and these marks occur in such places as to indicate that floating ice in large masses prevailed all over the higher portions of the northern and southern hemisphere.

During these conditions the country in many places was either permanently or temporarily under water, a condition which must have existed in order that icebergs should float from one district to another, and thus transport their masses of stone, gravel, &c. into regions from which they were far removed.

The effect of this severe climate seems to have been that various races of animals which had existed previously were exterminated, and left their bones in remote regions where they are now found: among these are the mammoth and other giants, which could not or did not survive the glacial period.

These are among the most marked peculiarities brought to notice by the indefatigable labours of geologists.

There are various methods by which we may proceed to assign a cause for a certain series of effects. First, we may *assume* some cause, and then examine whether this cause would explain the effects. But the cause thus assumed may rest entirely on its capability for explaining the known effects; and if it be a mere gratuitous assumption, it ceases to be satisfactory to the inquirer, for the reason that there is no independent evidence of its occurrence.

Under this head may be classed one of the suggested causes of the glacial epoch, viz. that the solar system travelling through space,* probably passes through strata of cold or of heat, which would produce great changes of climate on earth. This assumption, it is evident, is entirely gratuitous; we have no more evidence that the earth does pass through such variable strata than we have that the sun's light and heat are variable. In fact, we could explain more satisfactorily the changes of the glacial epoch if we were allowed to conclude that the sun ceased to give its light and heat for a certain period, and then gave it with its usual power.

If, however, the earth did pass through a stratum of cold sufficiently intense to produce the conditions of the glacial epoch, and then gradually emerged into a warmer stratum, there would be but one long winter during this period, which might extend over any number of centuries, but would be but one winter, without a summer.† There would consequently be only one period during which icebergs and masses of ice would float about the country; for it is of the utmost importance to remember that *cold alone* is not sufficient to produce floating icebergs, heat also being

---
* Poisson, *Theorie Mathémat. de la Chaleur.*
† See *Heat as a Mode of Motion*, by Professor Tyndall, p. 192.

requisite to liberate the ice from the localities where it has been formed.

When also we have evidence that the scratchings and other marks indicating the rapid transit of floating ice extend much higher up mountains than they do at present, it indicates that the heat must have been greater than at present, in order to melt the snow and ice at greater elevations; and this greater heat would have been impossible on the supposition that the earth was in a stratum of great cold.

Still, the assumption that the earth did pass through a stratum of cold, or that the sun's heat and light declined so as to produce the glacial period, has some adherents; therefore this supposed cause remains as one of the possible explanations of the observed facts, but which explanation may be compared with others given in subsequent pages.

The second cause to which we have referred as one assumed to account for the climate of the glacial period is the variable distance of the earth from the sun, technically called 'the excentricity of the earth's orbit.' This excentricity is such that it produces a variation of one and a half million of miles in opposite directions of the earth's mean distance from the sun. Now, as the proximity to the sun would cause a greater number of the sun's rays to fall on a given space than there would be if the sun were more distant, and as it is supposed that the sun's heat varies inversely as the squares of the distance, it became a subject of inquiry whether the excentricity might not formerly have been so great as to cause a far greater annual change of temperature than at present.

Sir C. Lyell, in his *Principles of Geology*, has treated this subject at considerable length, and has in addition pointed out a most important fact, viz. that although we are one-thirtieth nearer the sun in December than we are in June, a condition which ought to make our earth one-fifteenth colder in June than in December, yet when a comparison is made of the temperature of all places north and south, it is found that facts are entirely in opposition to this supposed law; for the *whole* planet is actually warmer in June when farthest away from the sun, than it is in December when nearest to it.

It needs only to ascend a mountain when the sun is nearly overhead, to convince us that although on the snow-covered peak we are probably 10,000 feet nearer the sun than in the valley below, yet we are colder, not warmer. Thus many other conditions are more powerful in producing heat than is our distance from the sun.

This second assumed cause may or may not produce changes in the climate of the earth. Whether we have at present data sufficiently accurate and definite to assign a value to the excentricity even 100,000 years ago, we think is open to question; but this assumed cause, like the preceding, may be weighed in comparison with those which will be subsequently given.

The third cause, to which we will briefly refer, is that so powerfully advocated by Sir C. Lyell in his geological works, viz. the influence of the distribution of land and sea in modifying and altering climate.

We have merely to examine the mean annual isothermal lines on any good map of the northern hemisphere, in order to perceive how much change is produced by the distribution of land and sea, and the action of currents in the ocean. For example, we find the mean annual temperature of 32° F. varies from latitude 61° on the west of North America down to 51° in Canada, extends to 66° in Iceland and to above 70° latitude north of Norway, and again comes down to 60° latitude in Russia. Thus England possesses a mean annual temperature of about 50° F., whilst portions of North America in the same latitude possess an average temperature of only 32° F., 18° less.

But what we see of the effects of the glacial epoch indicates that it was not the partial or local results of the elevation or depression of land which could produce these, for the effects are universal, over both northern and southern regions. The cause assigned seems feeble compared to these effects, and insufficient therefore to do more than explain minor changes and variations.

But it possesses the great strength of being in accordance with conditions now existing; for even now we find that elevations and submergences of land are occurring, and we know that great climatic differences do occur, according to the relative position of land and sea. For the land that is entirely surrounded by water is almost certain to possess a damp atmosphere, and one which may possess great cold or moderate heat, according as the ocean-currents come on this island from a northern or southern source. In our own islands we have the Gulf Stream, which, flowing from the warm latitudes of the West Indies, retains its heat for a considerable time, and warms our western and south-western coasts, and gives us a climate far warmer than is due to our latitude.

If, instead of this Gulf Stream, an ocean-current flowed from Spitzbergen, down east of Iceland and round the coasts of England and Scotland, then the British islands would have an excessively cold climate. But this effect would be only local, not universal over a hemisphere.

Thus whilst the meridian altitude of the sun at midsummer, and the length of time it remains above the horizon, are dependent upon the latitude of a place, and as the local heat is, *cæteris paribus*, dependent on the sun's altitude, yet this heat becomes considerably altered by such things as the existence of land north or south, or the flowing of ocean-currents from warmer or colder regions. The observations which have been made in various countries show that the climate does not depend *entirely* upon the altitude of the sun above the horizon. There are many modifying conditions, all of which must be considered.

If the air were always clear and bright, a very different temperature would prevail to that which results from a cloudy or foggy air. That the arrangement of land and water has something to do with climate cannot be doubted; but there may be *one* cause which produces both an excess of land or water on one part of the earth, and also a variability of climate.

Within the tropics there is a rainy and a dry season. In America the mean annual fall of rain is much greater than in Europe, and this would produce atmospheric conditions which must undoubtedly affect the climate.

The electric tension of the air is always changing; and there is no doubt that storms of rain and of wind, as well as of lightning, are the result of changes in the electric state of the air. When the air is clear, there is almost always an indication in it of positive electricity.

Thus, when heat or cold are spoken of in connection with climate, we must not assume a rigid law connected with the sun's altitude alone, but we must consider those other and more subtle influences which are at work, and which will produce a due effect.

Still, to obtain the effects of a tropical climate without at the same time having the sun's altitude considerable, or to obtain the effects of an arctic climate in middle latitudes whilst the sun still shone upon the earth as at present, is indeed drawing largely upon the elastic portion of science, and demanding an amount of concession in an inquirer, which will in the present day be rarely found.

Let us give all due weight to trade-winds which might have blown in former ages; to cloudy, misty, humid atmospheres which might have existed; to vast bodies of water which possibly surrounded the present inland countries; to variations in the electric and magnetic conditions of the earth which we cannot say did not take place; and still there will be the same necessity for some other universal and more powerful cause than a mere change in the relative position of land and sea.

From the well-known fact, that the altitude of the sun above the horizon is the main cause of heat in various latitudes, and that when the sun never rises above the horizon for many weeks or months we have an arctic climate, in localities thus affected, many inquirers have suggested that there must formerly have been something different in connection with the position of the earth's axis, which would cause the sun's altitude above the horizon to be such as to account for the singular variation of climate found during various past epochs.

The first attempted explanation of how such a thing might have happened was, assuming that two points which are called the North and South Poles of the earth were not formerly the extremities of the earth's axis, but that the axis of diurnal rotation occupied a different position *in* the earth to that which it does at present, and the North and South Poles consequently might have been situated somewhere nearer the equator than at present.

Thus instead of a warmer climate, or at least a moderate climate, prevailing in England, our islands may have been so near the former North Pole as to cause an arctic climate to prevail.

That this supposition should have received any approval from geologists seems singular, because if it had occurred, it would follow that two localities now on the equator must formerly have had a climate almost similar to an arctic climate, whilst a small portion of the northern hemisphere would have had an almost tropical climate.

But the facts of the glacial epoch indicate that the isothermal lines were very nearly parallel to the equator, and consequently lead to the conviction that the cold was nearly the same in all localities equidistant from the present pole of the earth.

The most powerful opposition, however, to this theory of the earth's axis of rotation having been situated differently to what it now is, was given by astronomers and mathematicians. Astronomers argued that during the history of astronomy all the evidence tended to prove that the earth's axis as regards its position in the earth was immovable. If it were not so, latitudes would be found to vary considerably, whereas our latitudes are constant. Therefore there was at present evidence of no change in the position of the earth's axis, and it was urged we had no evidence that any change had ever occurred.

Mathematicians also opposed this speculation, and affirmed that it was opposed to a well-known law, viz. that a body always rotated round its shortest diameter; and as the earth was a spheroid, the greatest diameter being in equatorial regions, it would be impossible for the earth to rotate round any other axis than that round which it now rotates.

An endeavour was made to escape from this mathematical objection, by suggesting that although now the equatorial diameter is much greater than is the polar diameter, yet probably formerly the greatest diameter was not situated where it now is, but in some other part of the earth; and if such were the case, then another axis of rotation might be possible.

These two suppositions, however, are, like many of those already mentioned, entirely gratuitous. They are not based on any observed facts or changes now in operation; and if they occurred, they would not account for the facts of which they are, the supposed explanation. For although a few minor variations occur as regards the effects of the glacial epoch, all the main evidence tends to show that an equal cold extended to all localities equally distant from the present pole of the earth, consequently the evidence proves that the pole of the earth was during the glacial epoch where it now is, and the axis consequently where it now is.

This supposition therefore is satisfactory or the reverse according as the inquirer looks at it superficially, or examines it carefully and compares it with recorded observations. As an abstract question, this theory may be stated to consist of two suppositions, which if they were true would have produced conditions which would not account for the facts for which they were supposed to account, and as such they have not many adherents.

Another cause of an astronomical nature was one that some sixty years ago attracted considerable attention, and for a time it was thought probable that it might explain the facts of geology.

It was found that when modern observations in astronomy were compared with ancient records, a continual decrease was taking place in the angular distance that the sun travelled annually north and south of the equator. This decrease, spoken of in astronomical language, was 'a diminution in the obliquity of the ecliptic.' The rate of this diminution was a slow one, 45" per century only, consequently 10° would, *if the rate were uniform*, occupy a period of above one hundred and thirty-three thousand years. It was not, however, this long period of time which caused farther inquiry in this direction to be stopped; but upon the subject being referred to M. Laplace, the then greatest and most popular mathematician of the day, he investigated a problem which he seemed to conclude embraced every phase of this subject, and came to a conclusion that the changes known to have occurred in modern times were confined within very small limits, and could not therefore formerly have ever caused such a change as would be powerful enough to explain the facts of geology.

The great reputation of M. Laplace as a mathematician and his known talent were so powerful as influences, that inquiry in this direction was entirely stopped; the subject was supposed to be exhausted, and any future investigator in the same problem was considered to be opposing himself to the well-known laws of physical science; thus no further investigation of any consequence has been made in this particular problem since that of M. Laplace, except by M. Le Verrier in *L'Ecliptic et l'Equinox de* 1800.

The problem which M. Laplace set himself to investigate was:

To what extent the plane of the ecliptic could vary by the action of certain physical causes.

The result at which he arrived was, that the plane of the ecliptic (which is, the plane passed over by the earth each year) would vary from a mean position not more than 1° 21'.

This result having been published, it was at once assumed that the obliquity of the ecliptic could never vary more than this amount, and consequently that no changes of climate could from astronomical causes occur sufficient to account for geological facts.

Within the past fifteen years, M. Le Verrier reinvestigated the same problem as that which had occupied M. Laplace's attention, and he came to the conclusion that the variation might extend to as much as 4° 3', an amount which differed considerably from that of M. Laplace, but was still insufficient to produce such a change of climate as that indicated by the evidence of the glacial epoch of geology.

Two such names as those of M. Laplace and M. Le Verrier are sufficient to prevent all inquirers from venturing over the same ground which they have trodden, with the hope of discovering any novelty; and though these two mathematicians differ *inter se*, yet we may probably be safe in assuming that the plane of the earth's orbit does not vary, and has not varied, more than a few degrees during endless ages.

If, however, we draw from this fact the conclusion that the *obliquity* of the ecliptic cannot and has not varied more than a few degrees, we commit a great error, and hastily jump at a conclusion. First, we fail to note what is occurring with other planets, and we indicate that we believe the only possible cause of a change in what is called the obliquity, must be due to a change in the position of the plane of the ecliptic. This conclusion would be unsound, and would indicate a want of geometrical knowledge as regards the actual cause of the obliquity.

Thus the two astronomical suppositions, viz. that the earth's axis was formerly situated at a different position to that which it now occupies, and that the plane of the earth's orbit may have altered considerably, are both considered untenable on mathematical grounds. Also the first supposition does not suffice to explain the facts, on account of the changes that are evident not being explained, even granting that the North and South Poles were formerly situated in parts of the earth far removed from their present position.

The second supposition, relative to a considerable variation in the position of the plane of the ecliptic would to a very great extent explain the facts at least of the glacial epoch; but this explanation has been considered untenable in consequence of mathematicians having agreed that the plane of the ecliptic cannot vary more than a few degrees from the position it now occupies.

In consequence of the difficulties and obscurities surrounding this subject, we have had no re-investigation of the problem for many years, and hence the cause or date of the glacial epoch of geology remains as much a mystery as when Professor Agassiz and others first published their opinion that ice was the principal agent in producing the effects recognised as having occurred just previous to the existing state of things on earth.

Among the numerous suggestions which have been offered to account for the glacial epoch is, that at the date when this condition occurred there was land between England and America, and of such extent as to turn the Gulf Stream from the course it now follows. Then, instead of flowing on the coasts of England and Ireland, and thus rendering our climate more genial than are similar latitudes in America, it followed another course, and thus left Great

Britain in the cold; a condition supposed to explain the glacial epoch.

Whilst it is undoubtedly true that the relative position of land and sea does produce certain modifications on climate, yet it seems scarcely probable that the cessation of the Gulf Stream on the west coast of England or Ireland should explain the former enormous glaciers in the Alps, the intense cold and drift formerly occurring in Siberia, or the extensive drift and erratic phenomena in North America, in South America, and in New Zealand.

The fact that in Germany, Sweden, Poland, and Russia, blocks in abundance are scattered; that many of these are of comparatively local origin, whilst the majority come from the north; seems scarcely to meet with an explanation by assuming that the Gulf Stream at the period of the glacial epoch did not break on the coast of England.

Few, we believe, among those who have heard this supposed cause consider it either satisfactory or sufficient; and even those who have assigned it as a probability, remark that the great accumulation of erratic blocks seems due to some more general cause, since not only are the blocks scattered in great abundance over northern Europe in a manner to show their northern origin, but those which occur in the northern parts of America, apparently in equal abundance, also point to a similar origin.

Many other suggestions have been offered relative to a cause for the facts known to have occurred during the glacial epoch; but there are scarcely any more worth discussing. Formerly it was conjectured that a great wave, rising from some unknown cause in the northern regions, dashed all over the northern hemisphere, and carried everything before it; or that a comet came into collision with the earth, and so— how is not explained—produced the effect. The present knowledge of geology renders it unnecessary to reply to this conjecture. It seems, however, agreed that such universal effects as are exhibited by the glacial epoch require an universal cause to explain them in an adequate manner.

CHAPTER IV.

THE THREE PRINCIPAL MOVEMENTS OF THE EARTH DESCRIBED AND EXAMINED.

In order to understand even superficially the following pages, the definitions given below should be thoroughly well known:

1. The Equator is an imaginary plane dividing the earth into two hemispheres.
2. The Poles of the Earth are two imaginary points on the earth's surface, distant 90° from all parts of the equator.
3. The Axis of the Earth is an imaginary straight line joining the poles, and therefore passing through the centre of the earth.
4. The Poles of the Heavens are those points in the heavens towards which the axis of the earth is directed. They are 180° apart, and are called the North Pole and the South Pole of the heavens.
5. The Equinoctial is a great circle in the heavens formed by the production of the plane of the equator to the heavens.
6. The Ecliptic is the great circle traced by the sun annually among the fixed stars.

DEFINITIONS USED IN ASTRONOMY.

7. The plane of the ecliptic is the plane along which the earth annually travels, and is the same plane as that along which the sun appears to travel.

8. The poles of the ecliptic are two points in the heavens, 90° distant from all parts of the ecliptic.

9. The obliquity of the ecliptic is the angle at which the plane of the ecliptic is inclined to the equinoctial: this angle is always the exact measure of the angular distance between the pole of the ecliptic and the pole of the equinoctial. It is also the measure of the angular distance of the Arctic circles from the poles of the earth, and of the tropics from the equator.

Let P in the annexed figure be the pole

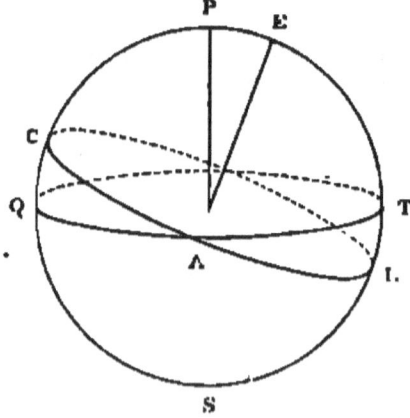

of the heavens; Q A T the equinoctial; E the pole of the ecliptic; C A L the plane of the ecliptic.

Then the angle C A Q will be the obliquity; and this angle is measured by the arc C Q, and is equal to P E.

For Q P = 90°, and C E = 90°, because the pole of the equinoctial is 90° from all parts of the equinoctial, and the pole of the ecliptic is 90° from the ecliptic; therefore C E = Q P. Take away the common part P C, and the remainder P E = C Q.

That is, the angular distance of the pole of the ecliptic from the pole of the heavens is always equal to the obliquity. At the present date it is about 23° 27′ 27″.

10. The equinoctial points are the points at which the ecliptic and equinoctial intersect, and the line joining these points is called the line of equinoxes.
11. The line of the solstices is a line joining those two portions of the ecliptic which are at the greatest angular distance from the equinoctial.
12. The declination of a star is its least angular distance from the equinoctial.
13. The polar distance of a star is the least angular distance from the pole, and is always the complement of the declination. Thus 90° − declination = polar distance.

DEFINITIONS USED IN ASTRONOMY.

14. The latitude of a star is the least angular distance of a star from the plane of the ecliptic, whilst the colatitude is the least angular distance of a star from the pole of the ecliptic.
15. The right ascension of a star is the angular distance (measured at the pole) of two meridians, one of which passes through the star, and the other through that point of the heavens at which the vernal equinox is situated.
16. Right ascension is counted from 0 to 360°, or from 0 hours to 24 hours, 15° being equal to 1 hour.
17. The longitude of a star is the angle formed at the pole of the ecliptic by two meridians, one of which passes through the star, the other through the vernal equinox.
18. In order to find the greatest and least meridian altitude of the sun at any locality during the year, proceed as follows:

Take the latitude from 90° and call the remainder the colatitude. Add the obliquity of the ecliptic to the colatitude for the greatest altitude of the sun, and subtract the obliquity from the colatitude for the least altitude of the sun. Example, required the greatest and least altitude of the sun during the year in latitude 51° 28′, when the obliquity was 23° 28′.

$$90° - 51°\ 28' = 38°\ 32' = \text{colatitude}.$$
$$38°\ 32' + 23°\ 28' = 62°\ 0'\ \text{greatest}.$$
$$38°\ 32' - 23°\ 28' = 15°\ 4'\ \text{least}.$$

When we refer to climate, this example will enable the reader better to understand the summer and winter changes due to a given obliquity. In bringing to the notice of the reader the subjects contained in the next chapters, we are compelled to treat them in a somewhat superficial and abstract manner. The results of many years' investigation and research cannot be fully given in a work of this description; several of them also are of a purely technical character, and would not therefore be understood by any but those whose course of study included the higher branches of geometry and mathematics; but as these investigations will be given in another work, those persons who are desirous of following the whole course of inquiry will be enabled to do so very shortly.

Quitting the subject of geology and geological evidence, we now turn our attention to that branch of astronomy which treats of the movements of our earth, and these movements will be examined as regards their geometrical changes. These changes can be investigated by the aid of geometrical laws well known to all who have studied geometry; and although some of the results arrived at have been obtained by long and laborious calculations, yet these results can be tested even by those who are not thoroughly skilled in geometrical analysis.

. . . .

The earth has three principal movements, two of which produce results noticeable annually to all observers. The first is its rotation on its axis; the second, its revolution round the sun.

The rotation of the earth on its axis is the cause of day and night. This rotation appears to be performed at an uniform rate, so that our day is of the same length that the day was thousands of years in the past. The rotation appears to be performed round an axis which is immovably fixed *in* the earth. The extremities of this axis, termed the poles of the earth, are consequently fixed on earth, and all localities on earth remain always at the same distance from these poles.

The revolution of the earth round the sun causes the change of seasons from summer to winter, because the axis of the earth is inclined at an angle to the path which the earth pursues. Thus, at one part of its course, the sun shines on localities from a point in the heavens nearly 47° higher than it does at the opposite period of the year; a fact which will be easily understood by the reader who examines the following diagram.

Figure 1 represents the position of the earth during December at the present time: the line joining N and s is the earth's axis; the angle formed by the equator with the ecliptic gives the extent of the tropics and the extent of the arctic circle. It will be seen that the sun is now vertical at all points distant from the equator as much as the angle formed by the equator and ecliptic. This angle at the present time is about 23° 27′ 27″. Thus the sun is vertical at all points south of the equator, whose latitudes are 23° 27′ 27″.

Six months after the earth was at fig. 1, it would have moved to the position fig. 2, which is the position of the earth in June. The sun is now vertical on the line marked Tropic of Cancer, which is 23°

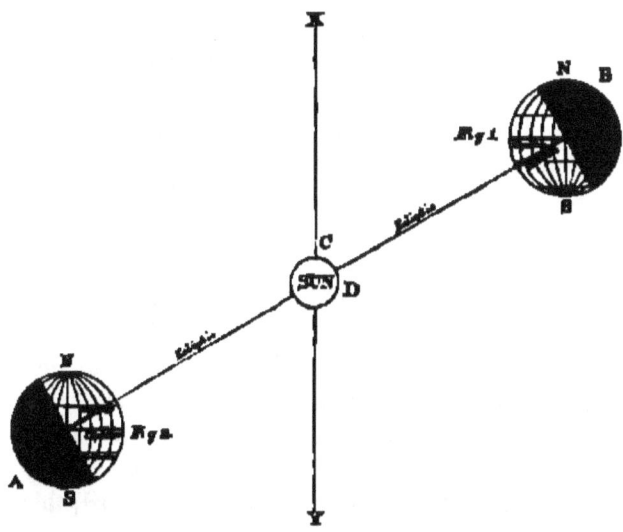

27′ 27″ north of the equator. Thus the sun has approached the zenith of all high northern latitudes twice 23° 27′ during its movement from fig. 1 to fig. 2. That is, its altitude has increased for midday by 46° 54′ from December to June.

The angle which the plane of the equator makes with the plane of the ecliptic is termed the obliquity, and it will be seen that this angle depends on the angle which the line N S makes with the plane of the ecliptic. Thus, as long as this line keep at a constant angle, the obliquity will be constant. This problem is of considerable importance, and it will be seen that the movement of the ecliptic alone will not cause any change in the obliquity unless the axis N S alter the angle which it makes with the ecliptic; and the axis cannot alter this angle without altering the distance of the pole of the heavens from the pole of the ecliptic. Thus, for example, we might turn this book upside down, and thus alter the angle which the plane of the ecliptic makes with the room, yet doing so would not alter the angle which N S makes with the ecliptic.

The third movement of the earth is one that can be noticed only when accurate observations in modern times are compared with those taken in previous years, because this movement is so slow that it cannot be discovered unless after long observation. This third motion is a change in *direction* of the earth's axis, whereby the whole earth is affected; this change in direction causing the earth's axis to point towards a different part of the heavens each year. Thus, while the axis is rigidly fixed as regards the earth in which it is situated, it changes its direction as regards external objects. The amount of change measured in angular distance is about 20" annually, or 1° in about one hundred and eighty years. It is this third movement of the earth which causes what is called the 'precession of the equinox,'—a difference between the solar and sidereal year,—a change in the declination and right ascension of celestial bodies, and many other changes requiring either repeated observation or elaborate calculations, in order to predict the true position of any celestial body at a given time.

The earth's axis by this continued movement traces a course in the heavens, which course is to the astronomer as defined and as marked as though a gigantic pencil had traced a course in the heavens, and had left its mark among the fixed stars. In modern times, when accurate instruments have been available, the course traced by the earth's axis is known to the greatest accuracy; whilst at more remote dates, when observations were less accurately made and instruments were less perfect, we can obtain close approximations only; yet by a combination of the two we may claim to know with great accuracy what is the course traced by the earth's axis in the heavens during the past eighteen hundred years.

A fourth movement of the earth is known to occur, but need here be only briefly mentioned, as it is compensated during a period of about nineteen years—that is, its cycle is completed during about that period; this movement is termed the nutation, and consists in a movement of the axis, by which it traces out a small ellipse, having a major axis of $18''\cdot5$, and a minor axis of $13''\cdot74$. This movement causes the earth's axis to trace a slightly wavy course in the heavens, and obliges us to separate the actual from the mean position which the axis occupies; but this is easily done, and but slightly complicates results.

In the future inquiries and explanations the *mean* position of the axis will always be referred to.

When it was accepted that the earth rotated on its axis and revolved round the sun, this change in direction of the earth's axis was described as a conical movement of the axis round the pole of the ecliptic as a centre. Thus the course traced by the axis was defined as an exact circle, the centre of which circle was the pole of the ecliptic.

When we remember how very imperfect were the observations of two hundred years ago, and how very slow and delicate is the curve traced in the heavens by the axis of the earth, it is evident that there must have been an insufficiency of data at the time when Copernicus assigned to the earth's axis so definite a movement as that of a circle, having for its centre the pole of the ecliptic. In nearly every modern work on astronomy, however, this same explanation is given, and we thus have the following as the accepted definition of this movement of the earth's axis:

'It is found then, that in virtue of the uniform part of the motion of the pole it describes a circle in the heavens around the pole of the ecliptic as a centre, keeping constantly at the same distance of 23° 28' from it in a direction from east to west.'

By the aid of a diagram we can better explain what the course is said to be that the earth's axis traces in the heavens.

E represents the position of the pole of the ecliptic, whilst the circle around E represents the course which the earth's axis was and is supposed to trace in the heavens. Thus P″ P′ P represent those positions to which the earth's axis was directed in remote ages, and if this be the true course traced, then some twelve thousand years ago the earth's axis was directed to a point near the star α Lyræ, and marked B on the diagram.

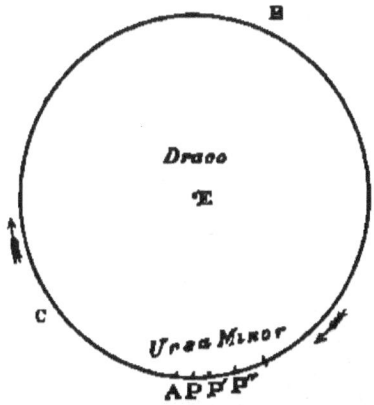

A Pole-star.   B α Lyræ.   C Stars of Ursa Major.
P″ P′ P Course traced by pole.   E Pole of ecliptic.

The point in the heavens towards which the earth's axis is directed is termed *the pole of the heavens;* therefore the circle shown above is the course of the pole of the heavens relative to the centre E.

This movement of the pole causes all stars that pass south of us at 6 P.M. on the 21st December to decrease their angular distance from the pole of the

CHANGE IN DIRECTION OF EARTH'S AXIS.

heavens about 20" per year. Of course all stars on the directly opposite side of the pole increase their angular distance 20" annually from the pole; those stars which in December pass the meridian about midday or at midnight scarcely alter their distance from the pole, on account of this point moving laterally as regards such stars. Thus we can ascertain in which direction the pole is moving by noting which stars decrease their distance most rapidly from the pole. The point E being the centre of the circle P" P' P, it follows that the angular distance of E from points in the circumference of this circle can never vary.

This movement can also be shown on the surface of a sphere by the annexed diagram.

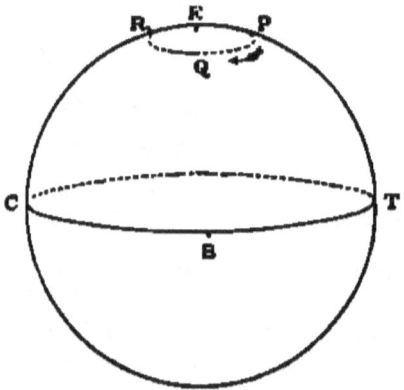

E represents the pole of the ecliptic, C B T the plane of the ecliptic, P Q R the course supposed to be traced by the earth's axis on the sphere of the heavens round the point E.

If P Q R be part of a circle having E for its centre, this circle cannot vary its distance from E. Neither will this circle vary its distance from the plane C B T. Thus E P, E Q, and E R will be equal; also P T, Q B, and R C will be equal.

It is not intended here to give a full description of all the results which follow this change in direction of the earth's axis, but merely to describe briefly the principal effects. One of these effects, viz. the precession of the equinoxes, may be understood by aid of the following diagram :

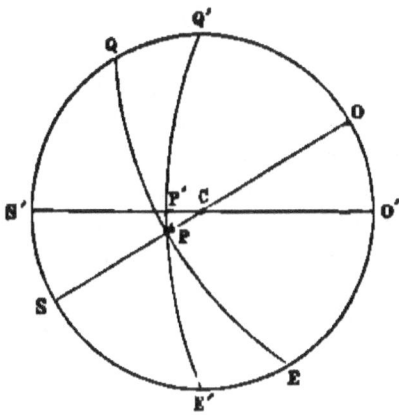

The circle S' S E' E O' O Q' Q represents the ecliptic, the pole of which is at C.

P represents the pole of the heavens at any date:

consequently P C is equal to the obliquity of the ecliptic.

The solsticial colure coincides always with the arc joining the two poles P and C. Therefore S P C O is a part of the solsticial colure.

The line of equinoxes is at right angles to the solsticial colure; thus, when the pole is at P, the line of equinoxes will be represented by the arc Q P E.

Let the pole P trace the arc P P' round C, the supposed centre of this arc. Then S' P' C O' will be the new position of the solsticial colure, and Q' P' E' the new position of the line of equinoxes.

As the earth moves round the ecliptic in its annual course from S' to S and to E', it follows that the equinox which did occur at E will, owing to the movement of the pole from P to P', occur at E'; therefore the equinox has preceded an entire revolution of the earth round the sun by the value of the arc E' E.

The value of the arc E'E for a polar movement of 20" is about 50". Hence the equinox occurs each year at a point 50" short of a complete circular course round the sun. As the earth passes over 360° in about 365¼ days, it follows that 50" occupy about 20 m. 19 s. Thus two successive vernal equinoxes occur 20 m. 19 s. before an entire circle round the sun is completed by the earth.

It will be seen from this description how important is this polar movement, for on it depends the amount of the precession and the different measures of time: for a complete revolution of the earth round the sun is completed in a sidereal year, whilst a tropical year is the interval of time between two successive vernal equinoxes. So important has this movement of the earth's axis been considered, that it has occupied the attention of the best mathematicians during more than a century. The system of investigation, however, must not be passed without notice; this system has been to give a physical cause why the earth's axis traces a circle round the pole of the ecliptic as a centre; and it has been agreed that the cause is due to the action of the sun and the moon on the earth's equator.

This theory was investigated by Frisi in *Theoria Geometrica Diurni motus*. M. D'Alembert treated of it in *Recherches sur la Précession des Equinoxes*, 1749. Lauden described the problem in *Mathematical Memoirs*; Euler, in the *Berlin Memoirs*; Walmesley in the *Philosophical Transactions*, vols. xlviii. and xlix.; Legrange in his *Memoir on the Libration of the Moon*; Vince in the *Philosophical Transactions*, 1787; M. La Place most elaborately in the *Mécanique Céleste*. Poisson, *Sur le Mouvement de la Terre autour de son Centre de Gravité*, gives a full description of this theory in the *Mémoires de l'Académie Royale*, 1829; and M. Poinsot in 1858.

When we perceive such an array of mathematical talent as the above devoted to the theoretical examination of this problem, we feel confident that any reader would at once assume, first, that the subject must be exhausted; secondly, that any writer who believed it possible to bring forward any new idea must be labouring under a delusion, or could not have examined the beautiful problems composed by those mathematicians whose names we have given.

Such a conclusion we must expect; for the theories are in themselves perfect, and no doubt ably explain why the axis of the earth must trace a circle round the pole of the ecliptic as a centre, *under certain conditions*. Also, if the earth's axis traced a circle round the pole of the ecliptic as a centre, no doubt the able explanations given by the above distinguished theorists would be perfectly satisfactory. There is, however, one important fact—a preliminary step only—which greatly deteriorates from the value of these most able and elaborate theories; and though it may not in the least interfere with their truth as theories, yet it is to the practical inquirer most important. This important fact is, that the earth's axis *does not* trace a circle round the pole of the ecliptic *as a centre*. Not only is this a fact demonstrated by the most simple of geometrical laws, but it can also be proved that during 1800 years, at least, the pole of the heavens has not traced any part of a circle round the pole of the ecliptic *as a centre*.

When, then, we find that the skilful theories invented to show a cause for the earth's axis tracing a circle round the pole of the ecliptic as a centre, have all been commenced by assuming that this is the movement of the axis, we can at once perceive that though true probably in themselves, they can have little to do with a movement not in accordance with the preliminary assumption on which they are based.

If, then, the pole of the heavens trace a circle round the pole of the ecliptic *as a centre*, and always maintain, as has been stated, the same distance from it of 23° 28', this work and our own labours are useless, and the elaborate theories referred to above are sound explanations of an actual fact. If, however, the pole of the heavens does not trace a circle round the pole of the ecliptic as a centre, then the theories referred to are at least questionable, as they profess to explain what does not occur.

Our duty at least is plain: we must demonstrate that the pole of the ecliptic is not the centre of the circle described by the earth's axis. If we accomplish this, we may fairly claim a hearing. If, however, it can be demonstrated that the pole of the ecliptic is this centre, then we must resign our opinion.

From the recorded observations of the earliest astronomers, we are able to ascertain that the change of direction of the earth's axis amounts to about 1° in 180 years; and this movement seems to be uniform in amount. Thus we have two conditions which must not be confused one with the other. We have an axis which is rigidly fixed in the earth, and as regards localities on earth never alters its position; but this axis changes its direction as regards external objects.

When we find that the change in direction of the

earth's axis explains in a satisfactory manner the precession of the equinoxes, and gives also a clear explanation of the variation in the duration of the solar and sidereal year; and when we also note how writer after writer has repeated the statement, that the pole of the heavens describes a circle around the pole of the ecliptic *as a centre*, and thus the two poles always maintain a constant angular distance of 23° 28' from each other, it seems almost unnecessary to attempt any researches in connection with this movement, with a hope of discovering any novelty or discrepancy.

If an examination resulted in bringing forward some probability, or even possibility, or in affording some other solution for the effects now explained by the change in direction of the earth's axis, such an examination would be at least interesting. When, however, we find that there are certain well-known geometrical laws which are violated if we assume one movement, but are consistent when we discover another slightly different; and when we also ascertain that the movement which is in accordance with geometrical laws enables us to obtain, by independent calculation, the value of certain astronomical data, hitherto only obtained by repeated observation; and when, also, we adopt a purely geometrical analysis as the base of our investigation, and obtain our results therefrom,—we believe it will be conceded that such an inquiry is worthy of considerable attention.

Under these circumstances we will venture to point out what there is unsatisfactory and untenable in one of the movements mentioned as that performed by the earth's axis; we will then explain the process which we adopted in the researches pursued for the purpose of solving this problem; and lastly, the singular results which were then brought to light, and their bearing upon the well-known facts of geology.

CHAPTER V.

EXAMINATION OF THE THIRD MOVEMENT OF THE EARTH, AND THE TRUE COURSE TRACED BY THE EARTH'S AXIS.

It is stated that the pole of the heavens describes a circle round the pole of the ecliptic as a centre, keeping always at the same distance of 23° 28′ from that centre.

It is also stated as the result of well-known observations, that from the earliest historical ages down to the present time there has been a continued decrease in the obliquity of the ecliptic to the amount of about 45″ per century.

These two statements to the superficial reader may present no discrepancy. To the geometrician they are flat contradictions the one of the other.

It is a geometrical law that the angular distance of the pole of the heavens from the pole of the ecliptic is the exact measure of the obliquity of the ecliptic. If, then, the pole of the ecliptic be the centre of the circle described by the pole of the heavens, no variation in the obliquity could occur, and thus the decrease recorded in the obliquity would be impossible. This decrease, however, is a fact so well ascertained, that it cannot be rejected, and we therefore have no alternative but to discard the statement that the pole of the heavens describes a circle round the pole of the ecliptic *as a centre*.

When we have satisfied ourselves that it is impossible that the pole of the ecliptic can be the centre of a circle, from the circumference of which circle it varies its angular distance, we have before us a problem or problems requiring the most careful and searching investigation. We find that a curve is traced in the heavens by the earth's axis. This curve is not and cannot be part of the circumference of a circle, the centre of which circle is the pole of the ecliptic; for if it were, then there could be no diminution in the value of the obliquity. Thus we have a curve traced by the axis which is undefined, for we at present know not the exact character of this curve.

There are several causes which may produce a change in the value of the obliquity, that is, in the angular distance of the pole of the heavens and the pole of the ecliptic. The pole, and hence the plane of the ecliptic, may slowly alter its position in the heavens, and either cause a diminution or increase in the angular distance between it and the pole of the heavens; or the curve traced by the pole of the heavens may be a circle, the centre of which is not the pole of the ecliptic; or the curve traced by the pole of the heavens may be an ellipse, or an irregular curve.

There can be only one condition under which there can be no change in the obliquity; and that is, if the pole of the ecliptic were the centre of a circle described by the pole of the heavens.

If the pole of the ecliptic were the exact centre of the circle described by the pole of the heavens, the following results would ensue:

1st. The curve traced by the pole would be exactly at right angles to the arc joining the pole of the Heavens and the pole of the ecliptic.

2d. There could be no variation in the obliquity of the ecliptic.

If the course of the pole were exactly at right angles to the arc joining the pole of the ecliptic and the pole of the heavens, it would be moving exactly towards those stars which, for the time being, were on the meridian of 0 hours right ascension, and away from those having twelve hours right ascension, and thus the course might be traced out with minute accuracy.

From some cause or other, the course of the pole cannot from recorded observation be traced with such accuracy as to assure us from the changes in North Polar distance of stars that it is moving exactly towards the meridian of 0 hours right ascension, consequently exactly at right angles to the arc joining the two poles, although it is evident that it is moving near that meridian; for, on comparing the declinations of various stars near the meridian of 0 hours right ascension in various catalogues separated by a long interval, we do not find that the greatest change is shown by those stars which were during the interval nearest the meridian of 0 hours right ascension, but it appears as if it were towards those stars which have for the time being about 352° of right ascension.*

From the fact that there is a decrease in the obliquity of the ecliptic, it follows that the pole of the heavens must decrease its angular distance from the pole of the ecliptic, and it could not decrease its angular distance from this pole if it moved at right angles to the arc joining the two poles, unless by one singular coincidence, nor could it decrease its distance if the pole of the ecliptic were the centre of the circle which is described by the pole of the heavens.

A decrease in the obliquity might occur if the pole of the ecliptic moved towards the pole of the heavens; but to do so would require that this pole followed the course of the other pole as it moved round the curve; and the instant the pole of the ecliptic changes its distance from the pole of the heavens, it proves that it is not the centre of the circle traced by the earth's axis.

If it be true that the change in star declinations is greater on a meridian of 352° than it is at 0° right ascension, it follows that this course of the pole must produce a decrease in the angular distance of

* This conclusion is arrived at from a most extensive research into all the known catalogues of stars from 140 A.D. to the present time.

the two poles, and hence in the obliquity; but we then have to define the curve traced by the pole, and describe its character as regards the pole of the ecliptic. It appears that when an investigation relative to the amount of change in the plane of the ecliptic was undertaken, and it was found that this change was possible,* no searching inquiry was made relative to the nature of the curve traced by the earth's axis. For it follows as a geometrical law, that even granting a considerable change in the position of the plane, and hence the pole, of the ecliptic, *still there would be no change in the obliquity of the ecliptic, if the pole of the ecliptic remained the centre of the circle described by the pole of the heavens;* consequently when it was decided that the plane of the ecliptic did vary, it followed as a law, that if a variation occurred in the obliquity, then the circle described by the pole of the heavens was deprived of a known centre, and was consequently left as an undefined curve. Whether the radius of this circle was greater or less than 23° 28' was not stated. That it could not be 23° 28' must have been known to all competent geometricians, yet the subject was then allowed to rest; whilst it appears as if some astronomers were satisfied to admit that by some singular coincidence the pole of the ecliptic, although it moved at least 45" per century towards the pole of the heavens, yet always managed to again shuffle itself exactly into the centre of the circle traced by

* See p. 91.

the pole of the heavens. Thus, although the distance of the two poles was formerly 23° 51' to 24°, and is now about 23° 27' 27", and consequently is a variable quantity, yet this variable quantity has been taken as the radius of a circle, and has been considered a constant quantity, and has been fixed at exactly 23° 28'; the pole of the ecliptic being always supposed the true centre of a circle from the circumference of which it was known to vary its distance.

It being a geometrical law that the obliquity of the ecliptic can be measured at any time by obtaining the angular distance between the pole of the heavens and the pole of the ecliptic, it also follows that we can obtain the angular distance of these two poles when we know the value of the obliquity; and as the value of the obliquity has been recorded during above 1700 years, we have records upon which we can base an investigation.

In these pages we deem it unnecessary to give in detail the various elaborate researches and calculations to which we devoted several years of labour; we will merely treat briefly that one which led us to results which are specially applicable to the glacial epoch of geology. It may be here stated, however, that the results as affecting certain problems in astronomy are not less important than is the one fact which specially explains the climatic changes indicated by geology.

On comparing the recorded observations relative

to the obliquity of the ecliptic, both in ancient and modern times, we must remember that observations made even 100 years ago were not as accurate as those now made with superior instruments. Observations made 1000 or 1700 years ago were even less accurate; and those ancient observations were never corrected for refraction, as refraction was then unknown. Thus, when we find that Ptolemy never gives the position of a star nearer than 10′, we know that the instrument he used—probably the astrolabe—did not read a less angle than ¼ of a degree; therefore any approximation to accuracy nearer than 10′ was almost impossible. We find, also, that the Greek astronomers made an error of 15′ in estimating the altitude of the pole* at Alexandra; therefore, when we require accurate records we must not refer to ancient observations. Before the use of telescopes, or the vernier, and before refraction was known, and observations were corrected for it, it is impossible that observations could be depended on.

When, then, we search for the records of the obliquity in past times, it is useless to depend much on those made before the use of telescopes, for these may be as much in error as was the latitude of Alexandra; therefore, if we go back 200 years, it is as far as we ought to go. We may, however, refer to the results obtained about 400 years ago, in order to note what appears to be an approximation

* Preface to Ptolemy's Catalogue, vol. xiii. of the *Memoirs of Royal Astronomical Society*.

at that date. The following is the value recorded for the obliquity in the years below mentioned:

A.D.	OBLIQUITY	OBSERVER
1437	23° 31′ 58″	Ulugh Beigh.
1500	23 29 52	Tycho Brahé.
1600	23 20 10	Hevelius.
1690	23 28 48	Flamstead.
1755	23 28 15	Bradley.
1795	23 27 57·6	Maskelyne.
1815	23 27 47·4	Bessel.
1825	23 27 41	Pearson.
1840	23 27 36·5	Airy.
1860	23 27 27·5	Nautical Almanac.
1870	23 27 22·2	″ ″

We may now inquire, what are these distances, and what they may enable us to ascertain.

In the first place, we know that the pole of the heavens moves at the rate of about 1° in 180 years, or if we refer this movement to a great circle, the pole will have moved about 6° between 1437 A.D. and 1870 A.D. The distances given above are the co-ordinates of *some* curve which the pole traces. This may be a regular or irregular curve, it may be part of a circle having a centre near the pole of the ecliptic, or it may be part of an ellipse. Although we may not be able to say what it is, yet we can at once declare what it is not. It is not part of a circle which has for its centre the pole of the ecliptic, because the radius, which is the value of the obliquity, varies as shown above.

To determine the nature of a curve, when we have the angular distances of several known points in this curve, from a point within the curve, is a problem quite within the powers of geometry to solve, and if we find that the curve which the pole of the heavens traces around the pole of the ecliptic be a regular curve, we can define the whole course of the pole during one revolution. If we can trace the course of the pole during one revolution, we can calculate its angular distance from the pole of the ecliptic by spherical trigonometry, and thus obtain by a geometrical calculation that which has hitherto been unattainable. The angular distance of the two poles is the exact measure of the obliquity of the ecliptic; so that if we can obtain the distances of these poles at any time, we can find the obliquity at that time.

Before attempting the solution of this important problem, we must consider well the results which would occur according as the pole of the ecliptic is a fixed or a movable point.

First: if it be a fixed point both as regards space and as regards the pole of the heavens, the curve traced by the pole of the heavens can, if a regular curve, be very easily defined.

Secondly: if the pole of the ecliptic change its position as regards space, and the pole of the heavens partake of this movement, and yet continue to trace a curve relative to the pole of the ecliptic, the conditions would (as regards these two poles) be exactly the same as if the pole of the ecliptic were at rest, whilst the pole of the heavens still traced round it a regular curve.

Thirdly: if during the regular movement of the pole of the heavens the pole of the ecliptic have also a movement not only as regards space, but as regards the pole of the heavens, it would be impossible, except by a most singular combination of chances, to give to the curve traced by the pole of the heavens any defined character. The pole of the ecliptic being the point to which we refer the co-ordinates, a change of position of this point would at once give us not a curve, but some very irregular course for that traced by the earth's axis on the heavens. In this investigation it would be desirable to obtain our results from the recorded observations of past time, instead of from those which have been made during the last few years, because we could then calculate for dates near the present time, and compare the results obtained by these calculations with the known observations now recorded. Thus we will endeavour to analyse the curve traced by the pole of the heavens by means of the records between 1437 A.D. and 1795 A.D., and although the observations may not be minutely accurate, still, by taking them as approximate, we may obtain an approximate result.

The annexed diagram will explain what the problem is which has to be investigated.

The curve A B C D E we suppose the curve traced by the earth's axis on the sphere of the heavens between the dates 1437 and 1795. Consequently A, B,

&c., will be the position of the pole of the heavens at certain dates, the length of the arcs A B, A C, &c., being ascertained by multiplying 20″·05 by the number of years between any two dates.

P represents the pole of the ecliptic; consequently A P will represent the value of the obliquity when the pole is at A, B P the obliquity when the pole is at B, and so on.

The question is, to decide whether A E is a regular curve traced on the surface of a sphere, and if so, what is this curve when referred to the point P. If it be not a regular curve, in what manner does it differ from one? In fact, to ascertain from the data supplied between 1437 A.D. and 1795 A.D. something definite about this curve; for at present to describe it as a part of a circle of which P is the centre is a geometrical absurdity.

We may, as a preliminary step, note some of the

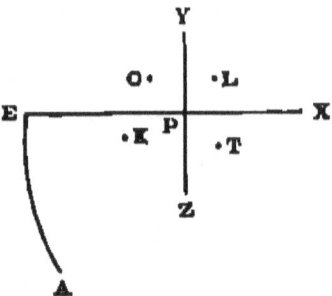

peculiarities of this curve. In the first place, as the arc is traced from A towards E, this arc approaches the point P. It therefore follows that if this arc be a part of the circumference of a circle, the centre of this circle must be on the opposite side of P to that on which the arc A E is located, and not between P and E; for if the centre were between E and P, the arc would increase its distance from P during its movement from A to E. Let A E represent the arc given above, P the pole of the ecliptic, which this arc approaches in its passage from A to E. Through P draw Y Z at right angles to E P X, and we then have four quadrants, viz. Y P E, E P Z, Z P X, and X P Y, in which to look for our centre.

If the centre were in the quadrant Y P E, the arc in passing from A to E would increase its distance from P. Thus, suppose O the centre, then with radius O A, which is greater than P A, the distance O E, which would be equal to O A, would be greater than P A, and this would hold good if the centre were anywhere in the quadrant Y P E.

In like manner if the centre were at L in the quadrant Y P X, P E would be greater than P A, unless the angle L P E were greater than the angle L P A.

If the centre were at K in the quadrant E P Z, P E would be greater than P A, unless K were situated near the line P Z, or was at a considerable distance from P, or made the angle E K P greater than A K P.

In the last quadrant X P Z, P A would be greater than P E if the centre were at T or near it, or were at some point near P Z, and considerably removed from P.

Thus we find that there are some positions where it is impossible the centre of this circle can be situated, whilst there are several other positions where it may be situated, in order that a decrease in the arc joining P with A E may occur as the curve is traced.

There can, however, be only one point for the centre in order that the distances P A, P E, and A E can be fulfilled. That the rate of decrease between P A and P E should be of a given value for a given length of arc A E, requires that one particular point only can be the centre of the circle of which A E is part of the circumference. Thus the curve which fulfils the conditions relative to its distances from P at various stations in this curve must be that one, and that one only, which is thereby indicated. The circumference of no two circles having different centres could be coincident over so long an arc as A E, neither would an ellipse, a parabola, or any other curve coincide with the arc A E as regards its minute details. If the arc A E be no part of a circle, it will be scarcely within the bounds of possibility that we can enable it to fulfil the conditions of a circle, by its approach to P in just the manner that a circle would approach it, unless A E is a portion of an ellipse having but a small difference between its major and minor axis. From a consideration of these and other similar conditions we may form a rough idea of what the curve may be, before we bring it to the test of analytical geometry.

Upon investigating this problem, we find that the curve here given is an arc of a circle having for its centre a point 6° from the pole of the ecliptic, and so situated that when the pole of the heavens has moved on in this curve to the date 2298 A.D., then the pole of the heavens, the pole of the ecliptic, and the centre of the circle described by the pole of the heavens will be on the same great circle of the sphere; or in other words, a meridian passing from the centre of this circle through the pole of the ecliptic will pass through the pole of the heavens at the date 2298 A.D.

Thus let E represent the pole of the ecliptic, s the position of the pole of the heavens, at the date 2298 A.D. Then a line drawn from s through E, and produced 6° to c, will give c the centre of the circle s m n, which is the circle described by the pole of the heavens, relatively to E the pole of the ecliptic.

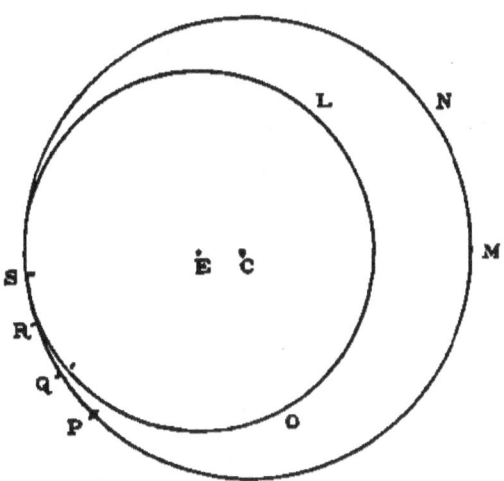

A circle of which E is the centre would be the circle S O L, all parts of which circle are equidistant from E, the centre. It is evident that the circle S M N is a much larger circle than is S O L, and various parts of this circle are at a much greater angular distance from E than are the points P Q R and S.

It will be evident that if the circle S P M N be the course traced by the earth's axis round the pole of the ecliptic, the pole in moving from P round towards S will gradually decrease its angular distance from E; but as this angular distance decreases, so does the obliquity of the ecliptic decrease: hence the angular distance P E will be the measure of the obliquity when the pole was at P; Q E will be the measure of the obliquity when the pole was at Q, and so on.

If, then, it be true that the circle S P M N be the course of the pole in the heavens, we shall be able to arrive, by a purely geometrical calculation independent of all observations, at the value of the obliquity for any date in the past or future; for it will be merely calculating the length of the arc E P, E R, &c., when other items such as the date are given. If we find that this test can be met, we may be tolerably certain that the curve here given is that actually traced by the earth's axis in the heavens relative to the pole of the ecliptic, and we may then follow it out in all its details.

The annexed diagram represents a portion of the surface of a sphere, the pole of which is C.

K

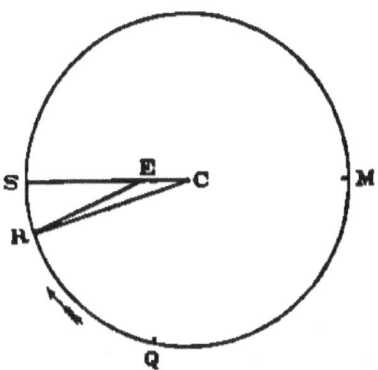

Q M S R is a circle on that sphere, and is the circle described by the pole of the heavens.

E is the pole of the ecliptic, distant 6° from c.

S the pole of the heavens at the date 2298 A.D., at which time the angular distance of s E will be 23° 25′ 47″.

From these items we can find by a rigid geometrical calculation the value of the obliquity for any date, provided we are correct as regards the course traced by the earth's axis relative to the pole of the ecliptic.

Let us calculate what the obliquity would be for the year 1871.

Let R be the position of the pole of the heavens at the date 1871, then the arc s R represents the value in degrees and minutes of the polar movement between 1871 and 2298 A.D., that is, during 427 years. The rate of the movement being 20″·05 per annum, we obtain 2° 22′ 46″ for s r. e c is known to be 6°, therefore s c = 29° 25′ 47″; and as c is the centre of the circle, then c s = c r, both being radii. We then have c s, c r, and s r, three sides of a spherical triangle, to find the angle at c.

Having found the angle at c, we can find the obliquity e r as follows:

e c r is a spherical triangle, of which we have two sides, e c = 6° and r c = 29° 25′ 47″, and the included angle at c, to find e r the third side.

To facilitate the calculations relative to the value of the obliquity, the process may be expressed in an algebraical form, so that the result may be readily obtained. The formula for finding the value of the obliquity at any date will be the following:

$$\frac{\{2298 - T\}\ 20''·05 = s.}{\sin.\ 29°\ 25'\ 47'' = \text{angle C.}}$$
$$\cos.\ C \tan.\ 6° = \tan.\ B.$$
$$\cos.\ O = \frac{\cos.\ 6° \left\{ \cos.\ (29°\ 25'\ 47'' - B) \right\}}{\cos.\ B.}$$

Where T represents the year (1871, for example) for which year the obliquity is required, o represents the obliquity.

The results here given, and the formula deduced therefrom, are obtained from an analysis of the curve traced by the earth's axis during about 358 years, viz. between 1437 A.D. and 1795 A.D. Now if this result be correct, we ought to be able to calculate what the obliquity should have been at 1840, 1850, 1870, &c., and ought to find that calculations based on this curve being such as we found it, agree exactly with the present accurate observations.

At the present time the value of the obliquity for any date in the future cannot be obtained by a geometrical calculation, because up to the present time the true course of the pole has never been known; therefore an empirical rule of merely adding at the rate of $45''\cdot7$ per century, or a proportion for a part of a century, has been the only one available. Whether the *rate* at which the obliquity decreases be an increasing or a decreasing rate has not hitherto been known; hence to decide what the obliquity would be 150 years in advance has not been possible.

Suppose, for example, that Flamstead, in 1690, wished to predict what the obliquity would be at 1840 A.D. Not knowing the true course of the pole, he could only estimate what it would be by finding what the change had been; and on comparing his observation at 1690 with that of Tycho Brahè, 1590, he would have found a difference of $1'\ 4''$, that is $64''$ for a century. Consequently, between 1690 and 1840 there are 150 years, and at the above rate there would be $96''$, that is $1'\ 36''$, which taken away from $23°\ 28'\ 48''$ would give $23°\ 27'\ 12''$ for the obliquity at 1840: that is, $24''$ less than it was found to be.

If, however, Flamstead had traced out the curve which the pole follows, and had compared the observations made by Tycho and Ulugh Beigh with those which he made, he might have obtained the formula given above, and by substituting 1840 for τ, he would have found for the obliquity 23° 27' 36"·5, which agrees to one-tenth of a second with the recorded observations at that date. If by the same formula he worked out the value of the obliquity for 1871, he would have found it 23° 27' 22", whilst recorded observation gives it 23° 27' 21"·83, showing a difference of only seventeen-hundredths of a second. Thus, had Flamstead been confident of his own and of his predecessors' observations to within only a few seconds, he might have made a discovery relative to the course of the pole which would have enabled him to predict the value of the obliquity at a date 181 years in advance, not with an error of nearly half a minute, but to within about one-tenth of a second.

Having from the recorded observations of the past found the course traced by the earth's axis, we will now submit this course to the severe test of comparing results obtained by calculation with those recorded as obtained by modern observations; and we find that during no less than 76 years we never differ even one second from recorded observations; a fact which almost amounts to a demonstration. Thus:

DATE.	OBLIQUITY BY CALCULATION.	BY OBSERVATION.
1871	23° 27' 22"	23° 27' 21"·83.
1840	23° 27' 36"·5	23° 27' 36"·5.
1825	23° 27' 43"·6	23° 27' 41".
1815	23° 27' 47"·5	23° 27' 47"·46.
1795	23° 27' 58"·3	23° 27' 57"·6.

This agreement is not merely for these dates; but when we go back some centuries, we find so close an approximation, that the curve seems demonstrated beyond a doubt, thus:

DATE.	OBLIQUITY BY CALCULATION.	BY OBSERVATION.
1690	23° 28′ 47″	23° 28′ 48″.
1500	23° 29′ 58″	23° 29′ 52″.
1437	23° 32′ 12″	23° 31′ 58″.

Thus, during nearly three hundred years, the calculation gives results which differ only a fraction of a second from modern and accurate observations, rarely ever more than a second from the observations of one or two centuries ago; and in fact these results differ less from the best observations than the best observers differed among themselves when making observations at the time.

Memos. of calculations relative to obliquity under the following conditions, and each calculation worked out independently of any other.

Radius of circle traced by earth's axis 28° 25′ 47″.
Angular distance of this centre from pole of ecliptic 6°.

Date at which pole of ecliptic, pole of heavens, and centre P of circle will be on the same meridian of longitude, 2208 A.D.

Then in the spherical triangle P′ C P we have P′ C = P C = 28° 25′ 47″, C being the centre of the circle of which P′ P is a portion of the circumference, E the pole of the ecliptic, P the position of the pole of the heavens at the date (2208 A.D.), P E = ∴ to 23° 25′ 47″, P′ the position of the pole of the heavens at any date in the past.

Required to find E P′, the angular distance of the pole of the heavens from the pole of the ecliptic; that is, the obliquity.

PROOF OF THE POLAR CURVE. 135

Take the polar movement in arc = 20″·05 annually.

For 1871.

Years elapsed from 2298 = 427 years.
427 × 20″·05 = 8560″·35 = PP′ polar movement in arc.

$$\frac{8560''\cdot 35}{\sin. P'C} = \text{angle at } P' = 4° 50' 24''.$$

P′C = 20° 25′ 47″
CE = 6°   } give for P′E 23° 27′ 22″.
P′CP = 4° 50′ 22″

Corroborated by *Nautical Almanac*, 1871, 23° 27′ 21″·83.

For 1860.

Years elapsed from 2298 A.D. = 438 years. Angle at C = 4° 52′ 52″, gives for obliquity 23° 27′ 27″.

Corroborated by *Nautical Almanac*, 1860, 23° 27′ 27″·38.

For 1850.

Years elapsed from 2298 A.D. = 448. Angle at C = 5° 4′ 40″ gives for obliquity, 1850, 23° 27′ 31″·4.

Corroborated by *Nautical Almanac*, 23 27′ 31″·95.

For 1840.

Years elapsed, 458. Arc PP′ = 9182″·9. Angle at C = 5° 11′ 26″.
As before, with two sides and included angle, find the obliquity P′E, which by calculation is 23° 27′ 36″·5.

Corroborated by Professor Airy's observations in 1840, which gave by observation 23° 27′ 36″·5.

For 1825.

Years elapsed, 473. Arc PP′ = 9583″·65. Angle at C = 5° 58′ 26″·2.
As before, with two sides and included angle, find P′E obliquity, which by calculation is 23° 27′ 43″·6.

Corroborated by Dr. Pearson's observation in the year 1825, which is recorded as 23° 27′ 44″.

For 1815.

Years elapsed, 483. Arc PP′ = 9684″·15. Angle at C = 5° 28′ 29″.
As before, the side EP′ (the obliquity) can be found by calculation.
By calculation for 1815, 23° 27′ 47″·5. By Bessel's observation, 1815, 23° 27′ 47″·46.

136   THE LAST GLACIAL EPOCH OF GEOLOGY.

For 1795.

Years elapsed, 503. Arc P P' = 10085".15. Angle at C = 5° 42' 3".
Obliquity by calculation for 1795 = 23° 27' 58".8. 1795 obliquity by Maskelyne's observations = 23° 27' 57".6.

Considering that these are results obtained from calculations depending on the correctness of the position assigned to the pole no less than 440 years in advance, and on the curve traced by that pole, we believe such a calculation and such accuracy are hitherto unexampled in geometrical astronomy.

· For 1755.

Years elapsed, 543. Arc P P' = 10887".15. Angle at C = 0° 5' 43".
Obliquity by calculation = 23° 28' 18".4. By Bradley's observation, 1755, 23° 28' 15".

This is a most remarkable circumstance; for on examining the table of refraction in Bradley's *Fundamenta Astronomiæ* we found he used a correction of only 3' 31"'4 for refraction for a zenith distance of 75°, instead of about 3' 34"'8, and a refraction of 30"'6 for 28° zenith distance, instead of about 31". Hence his *observations*, if correct, when thus corrected for erroneous refractions, would give 1"'5 too little for the obliquity. Also his instruments gave 51° 28' 39"'6 for the altitude of pole: an error again of 1"'6. Thus his observations cannot be expected to coincide *exactly* with calculation; but allowing for these two errors, we agree to 1".

For 1744.

Cassini de Thury in 1744 records 23° 28' 26" for the obliquity.
1744, by formula, the obliquity should have been 23° 28' 27".
Hugh Beigh's observation in 1437 seems a remarkable one. He

found the obliquity 23° 51' 5*". By calculation, the obliquity should have been at that period only 13" more, viz. 23° 32' 11".

As regards the value of the obliquity, we have calculated this for the next three or four hundred years, as well as for short intervals of time near our own date.

Thus the obliquity for 1900 will be 23° 27' 9"·6, and for 1950 23° 26' 50"·3.

The amount of the decrease in the obliquity between 1840 and 1870 should be 14"·1, which gives 0"·47 for the mean annual rate.

Between 1870 and 1900 there will be a decrease of 12·6, which will give a mean rate of 0"·42.

Between 1900 and 1950 there will be a decrease of 19"·3, which gives a mean rate of 0"·386.

Strabo, Geminus, Tatius, Proclus, Abraham Abnesda, &c. give $\frac{1}{15}$ of the circle, that is 24°, for the obliquity at their time; whilst Ptolemy and Hipparchus give somewhere between $47\frac{2}{3}°$ and $47\frac{3}{4}°$ for the variation in the sun's altitude between winter and summer.

By calculation the obliquity should have been 23° 56' 47" about 380 A.D.

We thus find that the curve which is and has been traced by the earth's axis in the heavens has a definite character. We find that when the most severe and delicate test is given to it, it proves to be, not as has hitherto been stated, a circle having a radius of 23° 28', but a circle having a radius of 29° 25' 47". Most particular attention must be called to the fact, that there is here nothing new introduced in the form of supposition or conjecture, theory or speculation. We have not assumed that something may or might have occurred formerly which does not occur now; but we deal only with the recorded facts of the past four hundred years. We have before us a curve traced by the earth's axis in the heavens; this curve is well and clearly defined during 400 years; all the *effects* resulting from this curve are well known, and have been well examined for hundreds of years; so that there is no contradiction between recorded facts and any new theory. When, however, it is stated that this well and clearly marked arc is part of a circle, the centre of which is the pole of the ecliptic, we at once affirm that this description of the curve is incorrect, and that to state that it is part of a circle the centre of which is the pole of the ecliptic, is not a true definition of this arc.

Trifling as it may appear, yet the apparently slight difference thus revealed is the key to the solution not only of the mysterious climate of the glacial epoch, but also of the date and duration of that epoch, and also of certain astronomical* paradoxes hitherto unsolved.

We do not think that any special technical knowledge is required to enable the reader to feel confident that the curve traced by the pole is not a circle the centre of which is the pole of the ecliptic. Nor does it require any great amount of knowledge to see

* In our next work, the *Cause of the so-called Proper Motion of the Fixed Stars*, the astronomical results of the true polar motion will be demonstrated.

that this curve corresponds to a circle the centre of which is 6° from the ecliptic pole; because this latter fact has been demonstrated by the exact agreement of calculations based thereon with the recorded observations of 400 years at least, if not of 1800 years; an agreement which, as before mentioned, would be impossible, if the curve were anything else than what we have defined it to be. It will be seen that in tracing out the curve made by the earth's axis we have treated the pole of the ecliptic as a fixed point instead of a variable one, as it is now supposed to be. We have done so for three reasons.

First: from a most extensive investigation of star latitudes during 1600 years, we cannot find any *trustworthy* evidence* that the changes in latitude prove that the plane of the ecliptic does vary; but we find their supposed changes due to another cause.†

Secondly: if the plane of the ecliptic does vary as regards the sphere of the heavens, and the pole of the heavens traces a circle round the point 6° from this pole, and follows the course of the pole of the ecliptic, the conditions as regards these two poles would be exactly the same as if the pole of the ecliptic were at rest. Consequently, when treating a problem relative to the obliquity or angular distance of these two poles, it matters not whether the pole of the ecliptic does or does not change its position *as regards space.*

* See evidence on this subject in last chapter.
† Demonstrated in our next work.

Thirdly: we find that the most delicate geometrical formula gives us exactly the angular distance of these two poles at any date, when the curve which the pole of the heavens really seems to trace has been defined; and as such an exact agreement is beyond the bounds of probability by chance, it appears that, no matter whether the ecliptic does or does not vary, yet the pole of the heavens partakes of this movement and traces its great curve.

Thus, just as the different parts of a steam-engine keep in gear, although the plane of the ship's deck, below which is the engine, varies its position considerably as regards the horizon; so if theory absolutely require that the plane of the ecliptic should vary, yet it will not alter the truth of this curve, which must remain absolute and certain as long as the distance between the two poles is found to vary as it has done during 1800 years.

For the above reasons we have referred the pole of the heavens to the pole of the ecliptic, as if the latter were a fixed point, and the truth of the preceding curve is not lessened even if it be proved that the plane of the ecliptic does vary. But as we also find that the discordances in the assigned positions of stars—attributed mostly to changes in their latitudes by older astronomers, and in some instances to 'proper motions' by modern astronomers, are accounted for by this movement of the pole, we have dealt with this problem as it may be dealt with, without interfering with the supposed change in the plane of the ecliptic.

REVOLUTION OF THE EQUINOXES.

So important and interesting are the astronomical problems connected with this course of the pole, that it is with reluctance that we do not here treat of them;* but as these pages are principally devoted to the climatic effects which would result from this movement, and which seem to be so entirely corroborated by the researches of geologists, we will merely confine our remarks to a description of some of its principal changes.

The revolution of the equinoxes would occur during a period of about 25,800 years, if the earth's axis traced a circle round the pole of the ecliptic as a centre, and always maintained the same distance from it of 23° 28'. But it will take about 31,840 years for the pole of the heavens to trace the circle found to be that indicated by the present polar movement, because this circle has a longer radius than 23° 28'.

The most important result connected with this change in direction of the earth's axis is, that during one revolution of the pole of the heavens round the pole of the ecliptic there would be a variation of 12° in the obliquity of the ecliptic, that is, the obliquity which at the date 2298 A.D. will be 23° 25' 47" was, when the pole of the heavens was exactly at the opposite point in the circle, 12° more, viz. 35° 25' 47"; and this condition would have prevailed about 16,000 years previous to 2298 A.D., viz. at the date 13,702 B.C. The effects known as the precession &c. now occurring would be *almost* identical for periods of 40

* They will be described in our next work.

or 50 years, no matter whether the pole of the ecliptic were or were not the centre of polar motion. Of course there could be no variation in the obliquity if the pole of the ecliptic were the centre; but all the other known results of the polar motion would occur in nearly the same manner. The differences in the two results, however, slight as they are, become manifest after a few years, and fortunately supply us with an additional amount of evidence.

It will be evident, if we examine the diagram on page 128, how it is that the obliquity of the ecliptic varies to so large an amount as 12° during one revolution of the pole of the heavens round the centre of polar motion.

The radius of the circle described by the pole is found to be 29° 25' 47", and the centre of this circle is 6° from the pole of the ecliptic. Thus C S in diagram on p. 130 is 29° 25' 47", but E S will be less by 6°; consequently E S is only 23° 25' 47", and E S represents the value of the obliquity at the date 2298 A.D., because the angular distance of the two poles is the exact measure of the obliquity.

When the pole of the heavens was at M, then C M the radius was as before 29° 25' 47"; and C E being 6°, the angular distance E M of the two poles would amount to 35° 25' 47", which would be the measure of the obliquity when the pole of the heavens was at M.

It will be also apparent that the pole of the heavens, in its course from M round to R and S, gradually decreases its angular distance from E, the pole of the ecliptic, and thus causes the obliquity to decrease from 35°25'47" when the pole was at M, to 23°25'47" when the pole will be at s.

As the circle M R s would be traced by the pole during 31,840 years, the half-circle M to s would occupy about 15,920 years. Thus, during about 16,000 years there would be a decrease in the obliquity of 12°, a sufficient alteration to produce most singular climatic changes, as will be shown hereafter.

It will be evident that as the pole of the heavens approaches that part of its course marked R and s, the *annual* decrease in the obliquity will be less than it was when the pole was at a point such as Q. It will also be seen that after the date 2298 A.D. there will be an increase in the obliquity, in consequence of the continued movement of the pole round its circular course towards M.

So marked will be the difference in results due to this movement of the pole, that by these results it must be judged; for although the changes between this movement and that hitherto supposed to have occurred are very slight, yet when we take long periods of time, the results are very different. Not the least remarkable are the climatic changes produced; for whilst we should have no changes of climate from astronomical causes, if the pole of the ecliptic were the centre of polar motion, we should find very great changes by the polar movement, which is in strict accordance with recorded observation.

CHAPTER VI.

THE CLIMATE OF THE EARTH DURING 31,840 YEARS.

WE will now work backwards from the present date to that period when the pole of the heavens was at the position M in the circle which it appears to trace.

When at M, the pole would be distant from the pole of the ecliptic 35° 25′ 47″, consequently the obliquity of the ecliptic would be of this value. In order to comprehend the climatic changes resulting from this condition, we will describe the sun's apparent course in the heavens during one year at the date referred to, and also call attention to what must certainly have been the climatic changes then prevailing.

We will refer to the conditions which would have prevailed in England, say in 54° of north latitude, during various periods of the year.

Starting from the period of the autumnal equinox, about the 21st of September, we should, with an obliquity of 35° 25′ 47″, find the sun's altitude at midday exactly the same as at present, and for a latitude of 54° the altitude at midday would be 36°. The sun would also remain twelve hours above the horizon at this period of the year; consequently, if the condition and distribution of land and water were the same as at present, the climate during September would have been much the same as at present in the same latitude and during the same month.

As the earth revolved round the sun, and three months elapsed, the winter position of the earth would have been reached, and on the 21st of December the sun would have attained its greatest south declination of 35° 25′ 47″. Under such conditions its midday altitude on the 21st December, at the winter solstice, for a latitude of 54°, would be less than 1°, and the sun would be visible above the horizon for scarcely more than one hour out of the twenty-four. At this period of the year all latitudes from 55° up to the pole would be in darkness, and would not see the sun for several days. Thus portions of Cumberland, Northumberland, and the whole of Scotland would remain for several weeks without ever seeing the sun above the horizon, just as is the case now in Iceland, and in portions of Norway and Sweden, Lapland, and northern Russia. The actual effect of an obliquity of 35° 25′ 47″ is to bring the arctic circle down to a distance of 35° 25′ 47″ from the poles, and to therefore subject all localities within this distance to an arctic climate during the winter of each year.

All localities within 35° 25′ 47″ of the poles have a latitude of from about 54° 34′ to 90°; all places on earth, therefore, with a latitude greater than 54° 34′

L

were within the arctic circle at the date 13,700 B.C., if the circle described by the earth's axis be correctly represented.

Whilst we have localities thus situated within the arctic circle, we should also at the winter period have all localities even further south with less sunlight than at present; for the sun's altitude would be 12° *less* than at present, on account of its being 12° further to the south. Consequently latitudes of 45° would then have possessed only the amount of sunlight and sun-heat now possessed in winter by latitudes of 57°; whilst latitudes of 55° would have experienced in winter the climate that now is experienced in latitudes of 67°.

When we examine the map of the northern hemisphere, we find under the 67th parallel of latitude the following localities: Behring Straits, the north of North America, Melville Island, Greenland, Iceland, and the northern portion of Lapland, and Norway. These regions are now, in winter, ice-bound and covered with snow; vast icebergs form on their coasts and in the rivers of these countries, glaciers of gigantic size exist, every river and stream is in winter converted into ice; and why? Because the sun at the winter period never rises more than one or two degrees above the horizon during several weeks.

At the date 13,700 B.C., the same conditions appear to have prevailed down to about 54° of latitude during winter as regards the sun being only a few degrees above the horizon. We are, then, warranted in concluding that the same climate prevailed down to 54° of latitude as now exists in winter down to 67° of latitude.

Thus in the greater part of England and Wales, and in the whole of Scotland, icebergs of large size would be formed each winter, every river and stream would be frozen and blocked with ice, the whole country would be covered with a mantle of snow and ice, and those creatures which could neither migrate nor endure the cold of an arctic climate would be exterminated.

These would be the climatic conditions during the winters at about the date 13,700 B.C., and they are of a character so very different from those which now prevail, that these must have left their mark on the earth's surface.

We have, however, only at present described three months of the year, viz. from September to December, or from the autumnal equinox to the winter solstice, and during this period the earth has glided into a climate, due to the sun's position relative to the earth's axis, of a most singular character as regards the northern hemisphere.

We will now describe the changes which would occur from the winter solstice to the vernal equinox, that is, from the 21st December to the 21st March.

As the earth left that part of its orbit at which the winter solstice occurred, and reached that point at which the vernal equinox took place, so would the meridian altitude of the sun increase from day to day until it reached 36° for the latitude of 54°. Thus a daily change would occur in the temperature, and a gradual increase from severe cold to temperate.

From the vernal equinox the earth would move on in its orbit till it reached the summer solstice, and at this date another marked change from present conditions would occur.

At the summer solstice the midday altitude of the sun for the latitude 54° would be about $71\frac{1}{2}°$; an altitude equal to that which the sun now attains in the south of Italy, the south of Spain, and in all localities having a latitude of about 40°.

There would, however, be this singular difference from present conditions, that in latitude 54° the sun at the period of the summer solstice would remain the whole twenty-four hours above the horizon; a fact which would give extreme heat to those very regions which, six months previously, had been subjected to an arctic cold. Not only would this greatly increased heat prevail in the latitude of 54°, but the sun's altitude would be 12° greater at midday in midsummer, and also 12° greater at midnight in high northern latitudes, than it ever attains now; consequently the heat would be far greater than at present, and high northern regions, even around the pole itself, would be subjected to a heat during summer far greater than any which now ever exists in those localities. The natural consequence would be, that the icebergs and ice which had during the severe winter accumulated in high latitudes, would be rapidly thawed by this heat; icebergs far exceeding in number those now annually liberated by the summer sun would be freed from even round the pole, and would float to more southerly regions; streams that had been frozen and blocked with masses of ice would be thawed; whilst the accumulated snow, ice, and frozen streams being thus rapidly and completely thawed, would produce floods which would inundate the country in all directions. A rapid thaw occurring in the present day after a few days' accumulation of snow and frost produces floods: what the effects would be of an almost tropical heat coming suddenly on the frozen masses and glaciers of northern Lapland we can easily imagine.

From the period of the summer solstice to the autumnal equinox the sun's meridian altitude would decrease from $71\frac{1}{2}°$ to $36°$ for a latitude of $54°$, and the changes of one year would have been passed through.

Thus during each year there would have been a fierce oscillation of climate in both northern and southern hemispheres; each winter England and all localities with the same latitude would have experienced an arctic cold, during which the earth's surface would have been covered with a mantle of snow and ice. Then followed the summer, during which excessive heat visited the same localities, liberating icebergs, which would be floated over the country, carrying with them their freight of boulders, gravel, &c. The snow and frozen masses on the mountains would be loosened, and then these masses slipping over the rocks, would leave their scratchings and marks in localities where snow and ice are now rarely found in any quantities. Each winter the whole northern and southern hemispheres would be one mass of ice; each summer nearly the whole of the ice of each hemisphere would be melted and dispersed. These effects would neither be local nor partial; they would occur over the whole northern and southern hemispheres, and would be particularly marked in those latitudes which were then on the borders of the arctic circle—viz. from about 45° to 60° of latitude.

The annual changes here described are those which *must* occur when the pole of the heavens is about 35° from the pole of the ecliptic; and in the circle already delineated, this condition appears to have occurred when the pole was at M in the circle. As already mentioned, the pole would occupy about 31,860 years to complete its circle round its centre; and as the geometrical investigation already described gives the date 2298 for the least angular distance of the two poles, it follows that about 15,000 years previously, viz. at about 13,700 B.C., the two poles were at their greatest angular distance, and the climatic changes annually occurring were then at their height.

When, however, the two poles were even 30° apart, a considerable annual variation of climate would occur, and the poles would be 30° or upwards from each other during half the period occupied by the one pole in describing its circle round the other. Thus the conditions of alternate cold and heat would have annually occurred during a period of about 16,000 years,—a sufficiently long time to produce effects of so marked a character as to render it easy to discover these effects on the earth's surface even at the present time.

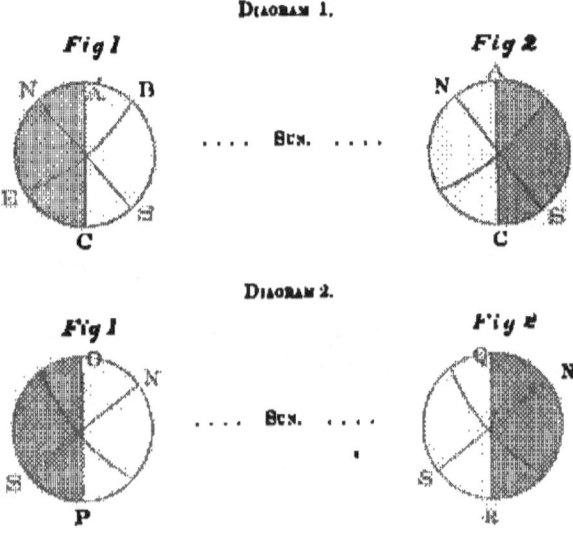

DIAGRAM 1.
Fig 1   *Fig 2*
.... SUN. ....

DIAGRAM 2.
Fig 1   *Fig 2*
.... SUN. ....

The change in direction of the earth's axis may be understood by reference to the above figures.

Referring to diagram 1, fig. 1 represents the earth at the period of the winter solstice, December at the present time; N s is the axis, and the dotted line directed to the sun is the plane of the ecliptic. The line A C shows the separation of sunlight from darkness on earth, and the small arc N A shows the extent of the arctic circle—now at midwinter all localities which are as near or nearer than A to N the pole will not see the sun for many days, and all such localities are within the arctic circle. At the present time this circle is distant from the pole about 23° 27' 30".

During a period of 16,000 years, the earth's axis having described half its circle by a conical motion, it was in the position shown by N s, figs. 1 and 2, diagram 2.

The winter solstice would now occur at fig. 2, 180° round the ecliptic from the position it occupied 16,000 years previously. The line Q R represents the line of sunlight on earth, the arc Q N the extent of the arctic circle. At the remote date referred to the extent of Q N appears to have been 35° 25' 47"; consequently all localities within that distance from the pole were within the arctic circle. As before pointed out, a locality 35° 25' from the pole has a latitude of 54° 35'; consequently the whole of Scotland and part of the north of England were within the arctic circle at the period referred to. Upon examining the climatic conditions which would prevail with an obliquity of 35° we must call attention to the marked difference of temperature between midsummer and midwinter in those latitudes now termed temperate, whilst in central portions of the earth there would be the same uniform temperature as at present. From the poles down to 54° of latitude we should in winter have had an arctic climate, which would naturally cause all our mountains in England and Scotland to be covered with vast masses of snow and ice, whilst the whole country would be frozen. All localities in the northern hemisphere would in winter only have a colder climate than at present—a climate similar to that which now prevails 12° more to their north. Consequently in mountains situated as are the Alps the winter snow-line would have come down much lower than at present; also the glaciers and moraines would have extended far lower than at present, even during the most severe winters of modern times. An altitude of the sun less by 12° than at present would produce most marked effects, which would be visible even at this date.

When we find that such must have been the condition produced by the earth's axis tracing the curve which enables us with minute accuracy to calculate the value of the obliquity at the present time, and to obtain by a calculation results hitherto only obtained by observation, it seems most noteworthy to find in such a work as Sir C. Lyell's *Principles of Geology*, p. 192, the following remarks relative to the Drift or Glacial Period:

'The temperature which prevailed in the valleys of the Thames, Somme, and Seine at the era in question was, according to Mr. Prestwich, 20° Fahrenheit colder than now, or such as would now belong to a country from 10° to 15° of latitude more to the north.'

The most accurate results obtained by geometry seem to agree exactly with the recorded facts of geology, and the same curve which explains clearly astronomical phenomena also explains certain geological facts.

Marked as are the facts relative to the extreme cold experienced in latitude 54°, we believe that the change to great heat each summer will also receive corroboration, for the effects of this heat must have left some traces which are visible.

The great heat of summer would be shown by two very marked effects. In consequence of this great heat, we ought to find that even in latitudes higher than at present there remains evidence of boulders, gravel, and drift having been carried formerly in far greater quantities than at present. Also, and this is most important, we ought to find in mountain regions such as the Alps, that whilst there was evidence of the glaciers having come *down* much lower than they now come, there was also evidence of the scratchings and markings of melted ice *far higher up* than what is now the line of perpetual snow. The fact that the sun's altitude in summer was 12° more than at present would cause the snow in summer to melt at much greater altitudes than it now does; and although much of this evidence must now necessarily be concealed by the perpetual snow, yet we might expect that some evidence would be found; and we have merely to search the works of various geologists who have studied glaciers to find that such a condition has been well known to them for a long period.

The effect of the sun's daily increasing altitude causing the heat to daily increase, we should find that the warm climate seemed to approach from the south with the sun; consequently the snow and ice would be melted in the south and on the south side of hills before it was on the north. The icebergs therefore would almost invariably drift southwards, because north of them there would still be ice and snow; consequently unless the formation of the country was opposed to it, the drift would always be in a southerly direction.

As remarked in preceding pages, the climate during the vernal and autumnal equinox ought to have been much as at present, because the sun's altitude was the same, but during the summer the heat would be great. Whilst the heat in the summer in the northern hemisphere would have been very great, the cold in the southern hemisphere would also have been great; consequently we should naturally expect that all animals that could do so would annually migrate, and thus avoid the extreme heat and cold which then oscillated on earth.

By comparing what is known of the present habits and powers of locomotion of animals with what may have occurred in former times, we may define not only what is possible but what is probable in the past.

Taking first the elephant. We have seen in Africa elephants travelling day after day at an average rate of about fifteen to twenty miles, browsing as they went, and merely migrating from one locality to another. Thus in three months these creatures might travel with ease from 1500 to 2000 miles—that is about 25° to 30° of latitude; consequently elephants whose habitat was India, in 30° N. latitude, might easily find their way to Siberia in three months, and in those regions would obtain a succulent vegetation, caused by the great heat, well suited to their taste.*

If the instinct of these animals was sufficient to tell them that in less than six months an arctic climate would prevail in that very Siberia, where in summer the sun was twenty-four hours above the horizon, and where its midday altitude was no less than 65°, they would start in June on their return journey. But if deluded by the heat, and lingering on till September, they then started on their return journey, they would be too late, and out-paced by the sun, which would sink each day far more rapidly than at present. Consequently these creatures, which in living species are dependent on water for their very existence, would be overtaken by an arctic

---

* This subject of migration during the Glacial Epoch has been largely and ably dealt with by Mr. Charles Darwin in *Origin of Species*, &c.

climate, which would rush down from the north with the first north wind, freeze every pool and river, and leave the elephant to die from thirst, and to deposit his bones and tusks in Siberia, to be found 17,000 years afterwards by men who, wondering at the vast facts accumulated around them, could yet find no explanation of these facts, in consequence of the stealthy movement of the earth's axis having so changed the climate as to conceal the causes which had led to these grand results.

If we assume that man was a resident on earth at that time, we can perceive that he was not likely to take up his residence in latitudes where he was frozen in winter, and exposed to twenty-four hours of sunlight and great heat in summer. To man in what we may call a primitive state, extreme cold is not so suitable or pleasant as warmth; thus all latitudes from about 50° up to the pole would, during the Glacial Epoch, have been unsuited to man, the cold of winter and the fierce heat of summer being alike inconvenient. Therefore we may fairly conclude that it is not in regions north of 54°, or even near it, that we shall find many remains of the antiquity of man; for if man then lived on earth he would undoubtedly have confined himself to a belt of about 30° on each side of the equator, and the daring hunter and pioneer alone would venture to journey north during a summer.

The winter in each hemisphere, when the obliquity was between 35° and 36°, would last with considerable severity during at least four months each year. During this winter the frost would prevail without intermission; every stream and river, therefore, would be frozen and pent up for four months, whilst there would be an accumulation of four months' snow on the earth's surface. When the streams were rapidly thawed, and the deep masses of snow became subjected to the heat of an almost vertical sun, the amount of icy water thus suddenly produced would cover the country in many places in such quantities as to form lakes of large size; and when we consider that each year during 16,000 years such a condition would prevail, we find that the forces to produce such results as are now known were amply sufficient.

Let us now note what evidence there is of such a condition having once prevailed, and we find the following in Darwin's *Origin of Species*:

'We have evidence of almost every conceivable kind, organic and inorganic, that within a very recent geological period central Europe and North America suffered under an arctic climate. The ruins of a house burnt by fire do not tell their tale more plainly than do the mountains of Scotland and Wales, with their scored flanks, polished surfaces, and perched boulders, of the icy streams with which their valleys were lately charged. So greatly has the climate of Europe changed, that in Northern Italy gigantic moraines, left by old glaciers, are now clothed by the vine and maize. Throughout a large part of the

United States, erratic boulders and rocks, scored by drifted icebergs and coast-ice, plainly reveal a former cold period.'

The irregular accumulation of drift materials and the peculiar adhesive character of the mud which adheres to the pebbles, &c., and also the absence in almost every instance of marine or fresh-water fossils, seem to indicate with great distinctness that melted snow or ice has been the chief agent in producing these effects.

From an investigation of the diagram, showing the course which it appears the earth's axis has traced, it will be seen that from the period 2298 A.D., counting back from 16,000 years previously, the obliquity decreased from 35° 25' 47" to 23° 25' 47". At 8000 years previously, that is at 24,000 years, counting from 2298 A.D., the obliquity would have been about 29° 25', and at this date the climatic changes would begin to be marked each year. Thus, although at the date 13,700 B.C. the last glacial epoch would have been at its height, on account of the annual changes of climate being then greater than they were either previously or subsequently; still, during about 8000 years before and during 8000 years after that date, there would be great annual change in the temperature, sufficient to produce much more ice and snow than at present, in low latitudes during winter, and much greater heat in summer, by which heat more icebergs would be liberated from high northern latitudes in summer, than is the case at present. Thus the largest boulders and the greatest effects would have occurred at the date 13,700 B.C.; but after this date there would have been an epoch of 8000 years, during which masses of frozen snow and ice would be annually floating over the earth's surface, and thus scratching and marking the boulders which had during the previous and most intense period of change been deposited in various localities.

Such evidence exists everywhere where geologists have carefully examined the details, and is particularly manifested about Lake Superior.

When we examine the actual results which must follow the fact of there being an angular distance of above $35\frac{1}{2}°$ between the two poles, viz. that of the ecliptic and of the heavens, we find that the arctic circle would extend down to $54\frac{1}{2}°$ of latitude, and hence a far more extended sheet of ice and snow then prevailed than is the case at present. Such being the certain result of the above conditions, we may compare these with the remarks of Professor Agassiz in his paper 'Erratic Phenomena.'

'Again (he says), if we take for a moment into consideration the immense extent of land covered by erratic phenomena, and view them as produced by drifted icebergs, we must acknowledge that the icebergs of the *present period* at least are insufficient to account for them, as they are limited to a narrower zone. And to bring icebergs in any way within the extent which would answer for the extent of the distribution of erratics, we must assume that the

CLIMATIC CONDITIONS.

northern ice-fields, from which these icebergs could be detached and float southwards, were much larger at the time they produced such extensive phenomena than they now are.* That is to say, we must assume an ice period; and if we look into the circumstances, we shall find that this ice period, to answer to the phenomena, should be nothing less than an extensive cap of ice upon both poles.'

It appears that if any writer had described what ought to have been the results of an obliquity of $35\frac{1}{2}°$, he could have given no more complete explanation than that offered by Professor Agassiz, and which is a description of the facts which he has observed.

We must now refer to another condition, which must follow the fact of an obliquity of $35\frac{1}{2}°$.

Whilst the arctic circle extended down to latitude $54\frac{1}{2}°$, the tropics would have extended up to $35\frac{1}{2}°$. Consequently all localities within $35\frac{1}{2}°$ of the equator would each year have had the sun in their zenith; but whilst at present localities just within the tropics, having a latitude of $23°$, would at their winter have an altitude of the sun at midday of about $43°$, latitudes just within the tropics, when the obliquity was $35\frac{1}{2}°$, would at midwinter have a midday altitude of the sun of only $19°$.

The effects of this condition ought to be clearly marked, for we ought not to find in the plains of the tropics any of those signs of boulders or erratic phe-

* Professor Agassiz does not seem to have noted the importance of heat to liberate icebergs. He speaks of cold as likely to do so.

nomena which we did find in northern and temperate latitudes. But on account of the sun's midwinter altitude having been only 19° at midday instead of 31°, as at present, for a latitude of 35½°, we should find that mountain ranges of considerable elevation showed signs of greater cold formerly than they now show, even when these mountain ranges are situated within 30° or so of the equator.

When we search for some evidence relative to this subject, we cannot over-estimate the value of the accurate observations of such a philosopher as Professor Agassiz, who in his 'Erratic Phenomena,' already referred to, gives the following description:

'Now that these phenomena have been observed extensively, we may derive also some instruction from the limits of their geographical extent. Let us see, therefore, where these polished, scratched, and furrowed rocks have been observed.

'In the first place, they occur everywhere in the north within certain limits of the arctics, and through the greater part of the temperate zone. They occur also in the southern hemisphere, within parallel limits; but in the plains of the tropics, and even in the warmer parts of the temperate zone, we find no trace of these phenomena, and nevertheless the action of currents could not be less there, and could not at any time have been less there, than in the colder climates. It is true similar phenomena occur in central Europe, and have been noticed in central Asia, and even in the Andes of South America; but these always in higher regions, at definite levels above the surface of the sea, everywhere indicating a connection between their extent and the colder temperature of the places over which they are traced.'

On comparing this geological evidence with that previously given, and with the facts that must result from an obliquity of $35\tfrac{1}{2}°$, it appears that the two sciences, astronomy and geology, agree in even minute details: astronomy indicating what ought to have occurred; and geological evidence proving that it did occur, and that the evidence is still visible.

The following description will explain the course of the pole of the heavens during one revolution round the pole of the ecliptic, and will show the date at which the value of the obliquity was most remarkable:

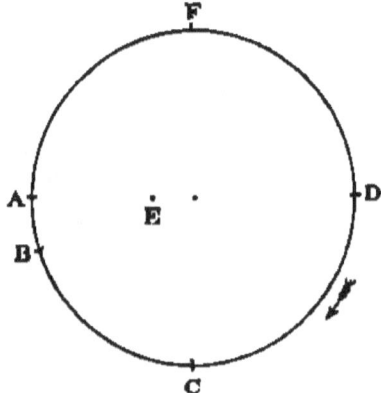

A will be the position of the pole of the heavens at the date 2298 A.D.; the arc A R, the nugular distance between A the pole of the heavens and B the pole of the ecliptic, will then be 23° 25′ 47″, which will be the value of the obliquity at that date.

B is the position of the pole about 2000 years ago; the arc B A is consequently the course of the pole during that interval; D E would be the obliquity about the date of the commencement of the Christian era, and would be about 24°.

C represents the position of the pole about the date 6000 B.C. This position is obtained as follows: The radius of the circle A B C D F is 29° 25′ 47″. The rate at which the pole travels is about 20″·05 per annum. Consequently, allowing for the size of this circle on the sphere, it will be found that it will take the pole 31,840 years to move round this circle.

During 7960 years, or one-fourth of the period, the pole will move over one-fourth of the circle, that is, from C to A. If, then, the pole be at A at the date 2298 A.D., then 7960 years previously it was at C, that is, at the date about 5660 B.C. At the date 5660 B.C. the angular distance E C would be about $29\frac{1}{2}°$, which would represent the value of the obliquity at that date.

D would be the position of the pole 15,800 years from the date 2298 A.D., that is, at the date

13,502 B.C., when the obliquity would be greatest, viz. 35° 25' 47", which is the angular distance of D the pole of the heavens from E the pole of the ecliptic at that date. When the pole was at D the alternations of climate would be greatest, and the Great Boulder period was at its height, the forces at work were then very powerful, as already described; the intense cold of winter in middle latitudes and the great heat of summer being most distinctly marked.

F shows the position of the pole when two-thirds of the circle has been traced, and when we reach a date about 21,000 years from 2298 A.D., that is, to the period of about 19,000 B.C. The pole then was again about $29\frac{1}{2}°$ from E, and the obliquity also $29\frac{1}{2}°$.

With an obliquity of $29\frac{1}{2}°$ the alternations between summer and winter in middle latitudes would be much greater than at present, and at this date the earth would have been gliding into its glacial epoch. It would reach its height when the pole was at D, and be gliding out of it when it reached C. Thus the conditions, as regards climate, would be suitable to produce all the observed effects during the period that it occupied the pole to move from F round to C, that is, during about 16,000 years. At a date about 31,840 years previous to the pole being at A, A should have been the position of the pole, and we go back to a date at which the obliquity would be only

23° 25′ 47″, *provided the circle which the pole traces be a concentric circle.* But as it appears that there is another motion in connection with the earth which would alter the results in very long periods such as 31,000 years, we will not deal with any further movements than those involved in one circular course of the pole of the heavens during 31,000 years, during which circular course there appear to be changes in the obliquity fully sufficient to explain the facts brought forward by geologists in connection with the glacial epoch of geology.

It may be here mentioned that the course which the pole appears to trace in remote ages is such as to give a repetition of a glacial epoch, though varying in many particulars from the last. Thus it would appear that the conclusions to which Professor Ramsey, Mr. Page, and one or two other geologists have arrived, seem to be corroborated by this movement of the earth's axis, which fully explains the phenomena occurring in astronomy.

The question whether a period of about 17,000 years, with the powerful forces at work resulting from an obliquity of $35\frac{1}{2}°$, would be sufficient to account for the facts of the glacial epoch is one requiring consideration. As before remarked, great results follow powerful causes in a short period, whilst a very long time is required to produce great results if the cause be feeble. With an obliquity of $35\frac{1}{2}°$ the causes would be very powerful, and we believe that those who understand what the climatic changes must be under such conditions will consider that 16,000 arctic or nearly arctic winters alternating with 16,000 tropical or nearly tropical summers, would be sufficient to explain all the observed effects of the drift and erratic phenomena. Whether or not another revolution of the pole of the heavens would be required to produce the effects, or several such revolutions, and consequently several successive periods of about 31,000 years, may be taken as a matter of detail; but we believe that one such revolution of the pole of the heavens during 31,000 years round the pole of the ecliptic, during which revolution the obliquity varied from 23° 25' to 35° 25', will be found sufficient to explain every fact of the glacial epoch.

The period spoken of in geology as the glacial period would have occurred between the dates about 6000 B.C. and about 22,000 B.C. At the date midway between these two periods would have been the height of the cold or glacial period. Hence from about 12,700 B.C. down to the present time we have the interval during which the alluvium and surface-soil formations took place; a space of time which probably some geologists may consider too brief, unless they bear in mind that far more powerful causes were then at work than are at present.

At the date about 22,700 B.C. we were just entering on the glacial period, therefore previous to this date we should have had a climate answering to an obliquity of about 29° to 23½°, and again to 29°; that is, while the pole was tracing the remaining half of its circle, viz. from c to A and F, this climate should have been but slightly dissimilar to that now prevailing on earth, and should not present the appearance of the great changes marked in the glacial epoch. This interval of 16,000 years would carry us back from the date 22,666 B.C., when the pole was at F, to 38,666 B.C., when the pole was at C, or near that point.

Whilst the pole was tracing this part of its course, and was approaching the conditions of the glacial epoch, the pliocene strata would have been deposited, and we should thus have, from the commencement of the pliocene down to the present time, about 40,000 years for these events. These results, however, are dependent upon whether one revolution of the pole of the heavens would be sufficient to account for the glacial epoch, or whether several such changes would be requisite; on this point we have found able geologists, to whom this problem has been shown, have different opinions, which probably may be fairly discussed and reasoned out at some future period.

Those geologists who advocate enormously long periods of time for every deposit must assume that former conditions were the same as at present; but as the very facts on which they reason indicate that the causes were often much more powerful than they now are, the time required for the production of the effects would be less. Thus during the Boulder period a shorter time would necessarily produce the same results when the forces were powerful.

When we find how such a movement of the pole

as that we have traced out gives during periods of about 31,000 years great changes of climate, we believe that the alternate beds of sandstone and shale, with coal, may find a solution; the coal being formed at one period and under one condition of the obliquity, the sandstone and shale under the other condition. Now as we find that in many of the coal-beds there are 70 and 80 layers of coal, we have at once a clue to the time occupied in forming these, if we take one revolution of the pole for each seam or layer; for to produce only 60 seams would require no less than two million years.

Between the carboniferous system and the boulder period a vast series of revolutions occurred, the number of which may probably be approximated to by those geologists who carefully study the fossils most likely to indicate past climatic conditions. We believe it quite within the powers of scientific investigation to be able to approximate very closely to the number of revolutions of the pole from the present time, through the glacial epoch, and on to the carboniferous system.

We have such distinct markings of the two climatic changes which occur during a revolution of the equinoxes when we examine the coal measures, alternating with sandstone and shale, that, as just stated, two million years would be required for these deposits. In the oolite system we have thin beds of limestone, beds of clay, Kimmeridge clay, Portland stone, &c., all indicating deposits under varied conditions. Above the oolite we have the green sand and chalk marl, very like drift materials, and chalks alternating with layers of flints. If the layers of flints each marked one revolution of the equinoxes, one great year in the world's history, one great period of 31,700 years, then, when we find masses of chalk, such as that forming Shakespeare's Cliff at Dover, or that of Flamborough Head, containing above 200 layers of flint, we find that 6,340,000 years would be required for these deposits. Above the chalk we have the eocene with its thick beds of clay and of limestone; above which is the miocene, with its limestone, fresh-water shells, and varied clays, each requiring many changes; then the pliocene, with its sand, clay, and pebbles; and then the diluvium, with its erratic blocks.

Thus, when we find it was 13,000 years ago that the last erratic boulder period could have occurred, it seems but as yesterday when we compare it with the vast epochs required for the cretaceous, oolite, and carboniferous systems, each of which can, we believe, be assigned with considerable accuracy a certain number of revolutions of the equinoxes.

We believe that when geologists free their beliefs from the mistaken idea that the pole of the ecliptic is the centre of polar motion, and thus that there can be no change, or very little change, in the obliquity, because there can be but little change in the plane of the cliptic, the various ages of the different strata will be worked out by a more accurate process than has hitherto been attempted; for the problem of the conversion of geological periods into astronomical time is solved, if the motion of the pole be such as is here demonstrated.

The reader, we consider, is now tolerably well acquainted with the general facts connected with the glacial epoch: he is also conversant with the climatic changes which must result from an obliquity of the ecliptic of 35° 25'. He has been able to form an opinion relative to the probability of the pole of the heavens having traced a circle round the pole of the ecliptic *as a centre*, or of its having traced a circle round a point 6° from the pole of the ecliptic. With these items before him, it becomes a question whether, when he reads a description of the known facts connected with the glacial epoch, he sees more clearly a cause for all the effects than he did when he believed the course of the pole in the heavens never produces, or has produced, any climatic changes on earth.

We have seen how the snow and ice accumulated during the winter in the northern hemisphere would be melted by the increased power of the summer sun at a date 13,000 years ago. We have seen how this rapidly-melted snow, leaving its icebergs, would form torrents exceeding many times any that now prevail. We have seen how the ice round the poles would be each year melted, or at least liberated, by the sun's heat, and sent floating over the northern hemisphere in quantities far exceeding any we now see. It is easy to comprehend how the rivers in all latitudes from about 45° up to the pole would each spring become roaring torrents, as down them poured the waters produced by three months of an arctic climate with its necessary accompaniment of snow. These are conditions which, it is not difficult to see, must result when the difference between the midsummer and midwinter meridian altitude of the sun was no less than 71°.

Having noted all these facts, let us turn to the admirably detailed account of the glacial epoch given by one of our most painstaking geologists, who so long ago was an advocate for the antiquity of man, and extract some few of the remarks from his paper 'On the Geology of the Deposits containing Flint Implements' in the *Philosophical Transactions*, 1864. Referring to climatic changes, excavations of valleys, Mr. Prestwich says: 'I have shown, on the authority both of continental and English geologists, as well as by the evidence brought forward by myself in this or in my former paper—

'1st, That certain beds of gravel at various levels follow the course of the present valleys, and have a direction of transport coincident with that of the present rivers.

'2d, That these beds contain in places land and fresh-water shells in a perfect and uninjured condition, and also the remains, sometimes entire, of land animals of various ages.

'3d, The extent and situation of some of these

beds of gravel so much above the existing valleys and river channels, combined with their organic remains, point to a former condition of things when such levels constituted the lowest ground over which the waters passed.

'4th, That the size and quantity of the débris afford evidence of great transporting power; whilst the presence of fine silt with land shells covering all the different gravel-beds, and running up the combes and capping the summits of some of the adjacent hills to far above the level of the highest of these beds, points to floods of extraordinary magnitude.

'These conditions taken as a whole are compatible only with the action of rivers flowing in the direction of the present rivers, and in operation before the existing valleys were excavated through the higher plains; of power and volume far greater than the present rivers, and dependent upon climatic causes distinct from those now prevailing in these latitudes. The size, power, and width of the old rivers is clearly evinced by the breadth of their channel and the coarseness and mass of their shingle-beds ; whilst the volume and power of the periodical inundations are proved by the great height to which the flood-silt has been carried above the ordinary old river levels—floods which swept down the land and marsh-shells, together with the remains of animals of the adjacent shores, and entombed them either in the coarser shingle of the main channel, or else in the former sediment deposited by the subsiding waters in the more sheltered positions. As the main channel was deepened from year to year by the scouring action of the rivers, the older shingle-banks were after a time left dry, except during floods, when they became covered up with the flood-silt, which extending also over the adjacent land and shores, was there deposited directly upon the rocky substratum. . . . . But not only have we these exhibitions of the power of the old rivers; it is further evident from the presence in the terrace-gravels of large blocks, often but little worn, and transported from considerable distances, together with much sharp and angular smaller débris, that there was some other power in operation besides the ordinary transporting power of water, great though it be. For the blocks in the one case would have shown an amount of wear in proportion to the length of transport, and the smaller débris would have been separated from the larger; whereas the blocks are always more or less angular, they are scattered indiscriminately through the gravel, are often associated with the most delicate and fragile shells and with bones of mammalia, but little or not at all worn. The only cause adequate to produce these results is, I conceive, the action of river-ice, whereby these blocks and a portion of the débris were carried down and deposited along the river channel, more especially in those parts where the currents may have been checked either by a widening of the river, or by the influx of a tributary stream. The recent phenomena with reference to the transport of blocks by ice on the St. Lawrence at its breaking up have been so well proved by Captain Bayfield and Sir W. Logan, and illustrated by Sir Charles Lyell and other geologists, that it is unnecessary to enlarge upon them here.

* * * * * *

'These results agree with and confirm the indications furnished by the organic remains, viz. that at the period of the high-level gravels the winter cold which so froze large rivers as to furnish ice-rafts capable of transporting innumerable boulders, many of five or ten tons weight, or more, for great distances, was not less than that of Moscow or Quebec at the present day, and that it may have been even lower. It is generally admitted that previous to this time, in the pliocene or early post-pliocene period the cold was still more severe. Then the greater part of England was under the sea; whereas Switzerland and the greater part of France had emerged from the sea at an earlier or miocene period, and there is no proof of their having been subsequently submerged.

'It was during this previous period of intense cold that the wonderful extension of the Alpine glaciers took place, and that many minor chains, such as the Jura and the Vosges, had also their glaciers. On the north of the Alps those old glaciers descended to within 1200 to 1000 feet of the present sea-level, whilst those now existing in Switzerland do not come lower than within 3400 feet of that level. M. Leblanc has calculated that such a difference of level might be accounted for by a reduction of the mean annual temperature of 12½° Fahr. But although that might give the limits to which glaciers could descend in these latitudes under ordinary circumstances and like conditions, it by no means proves that a greatly lower temperature may not have accompanied and hastened their enormous growth; nor when we look at the length and extent of the valleys, of which the fall is but small, over which the old glaciers passed, can their progress along surfaces so slightly inclined be compared with that where the inclination, as usually in their present beds, is steeper, and the channel narrower. I do not believe, therefore, that this estimate of M. Leblanc furnishes us with even an approximation to the extreme cold of that glacial period. . . . Although,' continues the same writer, ' I can conceive that granting an indefinite length of time, the wearing power of torrential rivers might effect considerable erosions, we shall find that other causes have assisted to produce the immense valley excavations we are now contemplating. For if the period is assumed to have been one of severe winter cold, we must follow out the consequences of that assumption not only with regard to the floods following upon the winter snows, but in all its collateral bearings.'

The reader who possesses the key to the polar movement, and realises the climatic conditions which must have resulted when the pole was 180° in its circle from where it now is, and when the arctic circle came down to 54½° latitude, cannot but admire the manner in which the various facts have been noted and the former conditions described in the above paper. Had the causes been known, and a writer had been asked to describe what the results should have been, he would have given these results in nearly the same words as are given above. When endeavours have been made to explain the effects of the boulder period, by assuming that the whole earth was subjected to one long uninterrupted winter, or condition of great cold, the actual facts which would have resulted if such an event had really occurred do not appear to have been quite sufficiently examined in detail, and compared with known effects.

If one long unbroken winter prevailed over the northern hemisphere, there would have been one long period during which that hemisphere would have been buried in snow and ice. When the winter gradually passed away, the snow and ice would gradually disappear. Such a condition, however, would not produce either large floating icebergs capable of transporting boulders nor masses of drift-ice; nor would the snow on the various mountains be melted, nor the drift and erratics scattered in all directions around mountains, &c.

The vast icebergs in arctic regions are liberated by *heat*, and the hotter the spring or summer the more numerous are the icebergs; and if an almost tropical heat were to now visit the North Pole, icebergs ten times as numerous as now would be floating down in middle latitudes; whereas, if winter and severe frost *always* prevailed in arctic regions, but very few icebergs would ever be seen in the Atlantic.

Also if *one* period only of cold visited the northern regions, only one scattering of detritus, and that a feeble one, could have occurred; and if this one period of cold were unbroken, fewer icebergs would have left arctic regions than now annually do so, because there was not sufficient heat to liberate them.

Well may we exclaim, then, with others, 'When we note the facts and hear the supposed explanation, we find this explanation feeble and insufficient.'

When we compare this idea of a long unbroken cold having lasted many centuries, and thus producing the glacial epoch, with the results following 17,000 winters of arctic cold down to 54° latitude, with 17,000 summers of great heat up to 70° latitude, we find that the facts seem to favour this movement of the pole. The heat of the summer which would scatter the ice round the poles *is as essential to explain the fact* as is the cold of the winter which produced the glaciers in Scotland and Wales.

Sufficient evidence has been adduced, we believe, to show that the movement of the pole which we have described in the preceding pages leaves very few facts connected with the glacial epoch unexplained; for it produces exactly those climatic changes which seem essential to yield all the effects which have been so long noted, and have as yet received no satisfactory explanation.

CHAPTER VII.

OBJECTIONS RAISED AND CONSIDERED.

ONE of the first steps to which any novelty should be subjected is criticism. The probabilities are considerably in favour of any new idea being easily disposed of; therefore, after having found that the polar movement to which we have referred was in accordance with recorded facts, that it enabled us to obtain by independent geometrical formula results hitherto only obtained by empirical formula, and that it explained geological phenomena hitherto unexplained, it appeared to have a tolerably firm foundation on which to rest.

When also we compared this movement with that assumed nearly three centuries ago, when it was not known that the pole of the heavens did vary its angular distance from the pole of the ecliptic, and when consequently the assumption that the pole of the ecliptic was the centre of the circle described by the pole of the heavens presented no geometrical impossibility, it appeared very much as if the former assumption had been made upon incomplete data.

In order to obtain all the objections that could be brought against our interpretation of the movement of the pole, we adopted four proceedings.

1st. We endeavoured to find some objection ourselves.

2d. We submitted the problem to those personal friends whose geometrical, mathematical, or geological knowledge qualified them to form opinions.

3d. We communicated with men of science both in England and America, and endeavoured to obtain from them such objections as we considered their special qualifications fitted them to offer.

4th. We read papers on this subject before those societies specially interested in this problem, and invited discussions on it.

Having adopted these proceedings during upwards of six years with only one result, viz. an absence of one sound objection against the truth of this problem, and being impressed with the conviction that it is of very great importance, and is the key to even much more than we have hitherto mentioned, we have now brought out the subject in a more popular form than it could be given in a scientific paper, and we will refer to, and answer, those objections which from their repetition appear to be those which recur to the majority of critics, and which will probably occur to the reader.

During the search for objections against this movement of the pole, we found from actual experience that a very peculiar opinion seems to exist with many persons relative to the value of *an objection*. It appears that when a novelty is submitted for criticism, any objection, no matter how erroneous or faulty this objection may be, is considered to entirely controvert the truth of the original problem. Thus, just as in former times men denied that the earth could be round, because if it were, people would be walking underneath with their heads downwards, like flies on the ceiling, so we have found several objectors offer remarks, which contained errors of so simple a nature as to be manifest to the merest tyro in geometry; yet these objections were considered unanswerable, until their error was pointed out; and as, in several instances, it was not considered worth while to point out the errors, the objectors most probably are still under the impression that their criticism was unanswerable.

The subject for discussion is one which may be brought within very narrow limits; it is—

What is the course that has been traced by the pole of the heavens during the past two or three hundred years as regards the pole of the ecliptic?

It has been hitherto taken as granted that this course is a circle, the centre of which circle is the pole of the ecliptic, and the radius of which circle is $23° 28'$, which is of course the angular distance of the pole of the heavens from the pole of the ecliptic, and is also the exact measure of the obliquity of the ecliptic. It is positively stated that the pole of the heavens never varies its distance from the pole of the ecliptic,* although it is admitted that the obliquity of the ecliptic varies as much as 45" per century.

This assumption is one to which we take exception, on the grounds that the value of the obliquity is always the same as the value of the angular distance of the pole of the heavens from the pole of the ecliptic; consequently, as the obliquity varies, the angular distance of the two poles varies, and hence the one pole cannot be the centre of the circle traced by the other pole. These two statements are geometrical facts which cannot be controverted.

It therefore follows that as there is known to be a decrease in the obliquity, it ought to be known that the pole of the ecliptic is not the centre of the circle traced by the pole of the heavens.

Unless we can define where the centre of a circle is located, and consequently what is the radius of this circle, the circle itself is undefined; as, therefore, it is a geometrical impossibility for the pole of the ecliptic to be the centre of the circle traced by the earth's axis, this centre must be somewhere else than coincident with the pole of the ecliptic; and until we can define where this centre is located, and hence what is the true nature of the curve traced by the earth's axis, we cannot claim to know what is the true course of the pole in the heavens.

When also it is stated that the pole of the ecliptic is the centre of a circle from the circumference of which circle this centre secularly varies its distance,

* Herschel's Outlines, Article 316.

a statement is made which is scarcely in accordance with that geometrical accuracy supposed and required to exist in so important a science as astronomy.

Among the first objections brought against the problem here described were those which are invariably urged by some persons against a novelty. They may be classed under the following heads:

1st. If this be true, why was it never discovered before?

2d. Surely the course of the pole must be accurately known now.

3d. If the course of the pole were not known, how could astronomical events be foretold as they are foretold with such great accuracy?

It is impossible anything new can now be brought out in astronomy.

No reply need be given to the first objection, for the same argument has been used invariably against every new truth, and may be urged against every novelty for the next thousand years.

As regards the second objection, viz. that the course of the pole must be known now, the reply is simple. The course of the pole is known: it is known to have been about 23° 29′ 1″ from the pole of the ecliptic at the date 1690; 23° 27′ 57″·6 in 1795; 23° 27′ 36″·5 in 1840; and 23° 27′ 21″·9 in 1871. It is known to decrease its distance from the pole of the ecliptic about 0″·45 per annum, and it is known to move in space about 20″·05 per annum. Therefore the course of the pole is known, and it is *upon a* knowledge of this course that our calculations are based. When, however, it is stated that this known course is part of a circle the centre of which circle is the pole of the ecliptic, an incorrect deduction is made from the known facts; for the course of the pole according to these observations is not a part of a circle having for its centre the pole of the ecliptic, but it is part of a circle having for its centre a point 6° from the pole of the ecliptic.

The third objection, viz. that it would be impossible to foretell astronomical phenomena if the course of the pole were not known, is one that is particularly attractive to many persons, and has by some been considered an insurmountable one.

From the reply to objection 2, it may already be seen that this has no real weight; for the course that the pole has traced during 300 years is known, and that which it will trace for 100 years to come is known *approximately*, and therefore, although it has not hitherto been known what this course really is, yet the facts resulting from it are known, at least for a few score years. When, however, we refer to long periods of time, the case is very different.

At present it is known that the obliquity decreases about 45″ per century, but whether this rate is decreasing or increasing has not as yet been known; although the recorded observations of the past, when corrected for those items unknown to olden astronomers, plainly show that *the rate is decreasing*.

For example, comparing even Flamstead's observations with those of Maskelyne, we obtain for 105 years a decrease formerly of 63''; which, even allowing for errors of observation, yet leaves a large margin beyond 45'' per century, the present rate. Again, comparing the recorded observations of Ptolemy with those of Ulugh Beigh, viz. between 140 A.D. and 1463 A.D., we obtain 83'' per century as the rate of decrease; whereas between 1840 and 1871 we obtain only 14''·6, which is at the rate of 47'' per century. It will be found during the next century that the rate of decrease of the obliquity is a decreasing one, and it will follow that either corrections will be required to be made and interpolated for some year, or the rate 47'' per century will have to be lowered. Thus, although observations may be predicted with *tolerable* accuracy for even a century in advance, a great variation is at once manifest when we refer to long periods in the past. Thus it is not known that the course traced by the pole of the heavens is one which, without any change whatever in the plane of the ecliptic, must have caused that pole to have been 35° 25' from the pole of the ecliptic at a date only about 14,000 years in the past. When, then, it is urged that phenomena cannot be predicted unless the true course of the pole be known, this argument is sound as regards long periods of time, but is not applicable when only one or two centuries are referred to.

The fact that the true course of the pole has not been accurately known has caused very many inquirers to imagine that the only possible cause of the variation in the obliquity was a variation in the plane of the ecliptic; and as M. Laplace concluded that $1°\ 21'$ was the limit, and M. Leverrier that it was $4°\ 52'$, it was considered that the obliquity could not vary more than double these values, whichever theorist was correct.

But both these theories were based upon investigations of a physical character, not of a purely geometrical nature; and whilst these views were accepted, it still appears that the singular impossibility of the pole of the ecliptic remaining the centre of the circle from the circumference of which it varied its distance did not attract the attention it deserved; nor does it appear to have been remarked that immediately we grant a variation in the obliquity, we have left the circle traced by the earth's axis without a centre. Also it appears that not sufficient notice was taken of the geometrical laws, that granting a change in the plane of the ecliptic, yet no change in the obliquity could occur if the pole of the ecliptic remained the centre of polar motion. And, granting a change in the position of the plane of the ecliptic *relative to the fixed stars*, still the course of the pole of the heavens *relative to the pole of the ecliptic* may be traced out, no matter how much or how little that plane varies.

Therefore, for all practical purposes relative to the value of the obliquity, we may refer the pole of the heavens to the pole of the ecliptic and obtain exactly the same results, no matter whether the plane of the ecliptic be movable or fixed in space, for the value of the obliquity is dependent alone on the angular distance of these two poles. The importance of this problem cannot be overrated.

When, then, it was concluded that the plane of the ecliptic could vary only 1° 21′, it was assumed that the obliquity could vary only 2° 42′; consequently, when geologists brought an array of undeniable facts to prove that climatic changes of a vast character had, not so long since, occurred on earth, and requested from astronomy some explanation, astronomical science was unable to afford such explanation, and in some instances endeavoured to avoid the puzzle by denying the accuracy of the geologist's facts.

It seems unfortunate that no geologist well acquainted with geometry had endeavoured to discover how a circle could have a centre from the circumference of which circle the centre was not equidistant. Some geologists, however, have even gone so far as to assert, that the inability of astronomy to explain facts, which must have been caused by an event of an astronomical nature, was admitting the incompleteness of astronomy as it at present stands.

There seems to have been a singular absence of geometrical investigation in connection with this movement of the pole of the heavens.

It is a geometrical law that a curve cannot approach a point at a rate that is uniform, provided the curve be traced uniformly. Thus it is a law that the pole of the heavens could not approach uniformly the pole of the ecliptic except under one most exceptional condition—viz. that the pole of the ecliptic always moved directly towards the pole of the heavens. (Even if this occurred, it must be remembered that the polar circle would still be undefined.) Yet in order to obtain the value of the obliquity for any remote date, it has been considered correct to multiply $45''$ by the number of centuries, and add the value thus obtained to the present value. Thus if the value were required for 14,000 B.C., the method was to multiply $45''$ by 140; and $1° 40' 30'' + 23° 28'$ would be given as the obliquity.

Another method was also adopted, which was alike remarkable for its absence of geometrical principle, and is a purely empirical formula, viz.

$$\theta = 23° 27' 54''\cdot 0 - 8''\cdot 4\text{n} \cdot 366\, t - 0''\cdot 000005\, t^2.$$
† denoting the years before or after 1800.

As the value of a formula mainly depends on its capability of being referred to a physical cause, that given above may be compared with the formula on page 131, which shows how one side of a spherical triangle indicates the value of the obliquity, and how this side may be calculated for all dates.

The reply, then, to the objection, that if the true course of the pole were not known astronomical events could not be predicted is, that the course of the pole is known for a few centuries with sufficient accuracy to predict astronomical events during that time. If, however, it is required to predict what the obliquity will be at the date 2298 A.D., this cannot be done at present by astronomers, except on the assumption that the obliquity decreases 45″ per century; and the obliquity for 2298 would therefore be given as 23° 24′ 12″, obtained by deducting 45″ for each century from 1795, whereas the obliquity for that date would be 23° 25′ 47″. Again, if it were required to define the obliquity at the date 14,000 B.C., only two statements could be made: either that there were no data on which this obliquity could be calculated, or, taking the decrease at 45″ per century, the obliquity ought to have been about $25\frac{1}{4}°$.

The conclusions from theory of M. Leverrier are, that the plane of the ecliptic may vary from the position it occupied at the date 1800 to the amount of 4° 3′, and that at the date 40,000 B.C. it did do so. He assigns for the excentricity at that date 0·0109; for the longitude of the perihelion, 28° 36′; and for the longitude of the node, 91° 59′.

Important as these theoretical calculations may be, they yet leave a far more important problem unnoticed; for they do not refer to the course traced by the pole of the heavens, on which the value of the obliquity almost entirely depends, but they leave this course as vague and as undefined as it was before; or they assume that the pole of the ecliptic is the centre of the circle traced by the pole of the heavens, under which conditions no change of the obliquity could occur, even granting a change of

4° 3′ in the position of the plane of the ecliptic as regards the fixed stars.

The fourth objection—viz. that it is impossible anything new can be brought out in astronomy—is an assumption which may be considered of value only by those who are incapable of judging of facts when placed before them. Whether or not a novelty which is true can or cannot be brought forward, is a subject not of opinion but of demonstration.

It was also urged that unless a physical cause, such as some action of gravity, could be assigned to the movement of the pole, it could not be accepted as true.

In reply to this objection, we state that the course of the pole is a matter of fact, just as is the rotation of the earth on its axis. The course which the pole traces is demonstrated by geometry to be a portion of a circle having for its centre a point 6° from the pole of the ecliptic; and even granting that we have not as yet given the cause of this course being traced, yet such absence of cause cannot invalidate the fact. We have as yet no cause assigned why the earth rotates on its axis in 23 h. 56 m. 4 s., nor why Jupiter rotates in only 9 h. 55 m. 49 s., nor why the moon rotates only once during its revolution round the earth; but to deny or ignore these facts, because we have no physical reason to assign for their occurrence is scarcely sound argument.

From objections of this weak character we must pass on to those with which we have been favoured by critics whose reputation or position was such as to give value to their remarks. We must, however, express our regret that many of these objections plainly indicated, that but little thought or care had been given to the problem which we had submitted to the authors of these objections, otherwise it is impossible their remarks could have been what they were.

Thus one writer, not unknown by repute, informed us that the course traced by the pole of the heavens was thoroughly known:

'That it was a circle around the pole of the ecliptic as a centre, from which centre it never varied its distance, remaining invariably at 23° 28' from this centre.'

This writer appeared either not to have looked at the problem which we had submitted for his criticism, or to be under the impression that, although we had devoted years of research to the question under discussion, we were yet unacquainted with the elementary but erroneous statement which he had but copied from a popular work on astronomy.

Another correspondent, also well known for his writings on popular astronomy, and to whom we had with great care pointed out that the change in direction of the earth's axis, amounting to 20"·05 per annum, was well known, and that the question was, what was the course thus traced by the axis, wrote to us as follows:

'I have carefully examined your problem, and

found it untenable. It is agreed by all mathematicians that the axis of the earth is rigidly fixed in the earth, and will remain so whilst the form of the earth is spheroidal. If the axis did change in the earth, all latitudes must change, which is known not to occur. Therefore your idea relative to the change of the earth's axis is not correct, as the earth's axis is immovable.'

This writer had started with a confusion of ideas relative to the change *in direction* of the axis, and a change *of the axis in the earth*. Hence his conclusions bore not only no reference to the problem, but he merely repeated a part of what we had written in our paper—viz. that whilst the axis was fixed in the earth, it was movable as regards space.

An objection of a far more valuable character was brought by a practical astronomer, viz. that 'The course traced by the pole of the heavens would not be at right angles to the arc joining the two poles; consequently, not directly towards the first point of Aries, except at the date 2298 A.D., but it would always be at right angles to the arc joining the pole of the heavens to the centre of polar motion.' Hence, he argued, the greatest change in north polar distance of stars ought not to be found in those having at the time 0 and 12 hours right ascension, but in those with about 23 h. 56 m. and 11 h. 56 m. right ascension. How is it, then, he inquired, that this result does not occur?

Having at the commencement of our investigation of this problem devoted several months to this particular point, we were naturally anxious to inform our correspondent of the result thereof, which was given as follows:

'Taking 20″·05 as the amount of polar movement per annum, we may approximate very closely to the amount to which each star will be affected by multiplying 20″·05 by the cosine of the star's right ascension. By adopting this process you will find that a star with even 12 m. right ascension will vary but little from 20″ per annum. When, however, we note that stars near the meridian of 0 hours and 12 hours right ascension have either so large a proper motion as to prevent this uniformity of change from being remarked, or *some other cause* is at work which is not known; yet it is a fact that a star's north polar distance does not vary with that uniformity which you have assumed. Thus the following stars vary as follows:

B.A.C.		RIGHT ASCENSION.			ANNUAL CHANGE IN NTH. POLAR DIST.
8314	Cephei	23h. 47m.	0s.		20″·14.
8319	Octantis, γ^2	23	49	0	19″·79.
8323	Tucanæ	23	49	0	20″·11.
8341	Cassiopeæ	23	54	0	20″·04.
8358	Ceti	23	56	0	20″·07.
8368	Piscium	23	57	0	20″·10.
	s Andromedæ	0	0	38	19″·03.
	ε Phœnicis	0	1	47	20″·04.
	22 Andromedæ	0	2	32	20″·03.
	γ Pegasi	0	5	30	20″·04.
	β Hydri	0	17	47	20″·28.

'From the above it is manifest that the pole may be moving in any direction, from 23 h. 47 m. R.A. to

17 m. R.A., as far as the decrease in north polar distance is concerned. I may add that I have compared the north polar differences of stars during long and short intervals, and by means of nearly every catalogue in existence. Comparisons have been made between Ptolemy's Catalogue, where I have converted his latitudes into polar distances, with those of Ulugh Beigh, Tycho Brahé, Hevelius, British Association Catalogue, Bradley's, Bessel's, La Caille's, Piazzi's, and those of modern times, and I can not only find no evidence clearly showing what you have assumed, but I find evidence indicating another and singular fact.

'Thus you are not correct in assuming that it is shown by the greater change in north polar distance of those stars which, for the time being, are on the meridian of 0 and 12 hours R.A. that the polar movement is directed *exactly* to those parts of the heavens; for the stars thus situated do not exhibit this marked difference from those within one or two degrees of them.'

The reader who will examine the diagram p. 130 will perceive that the movement of the pole will always be at right angles to the arc joining c with the pole, whilst it has been hitherto assumed to be at right angles to the arc joining E with the pole. At the present time the pole is situated relatively to E and c in the position R, and it is moving therefore at right angles to R c, and not at right angles to R E. The difference, it is manifest, is very slight, but it is still sufficient to account for many discrepancies now known to exist, when observations at long intervals are compared with each other.

Another objection was the following:

That the plane of the ecliptic is known to vary, in consequence of the increase in north latitude of the stars near the ecliptic in one direction, and their decrease in the opposite direction, and this change fully explains the decrease in the obliquity.

The following is our reply:

That it is most probable that the plane of the ecliptic does vary is granted; but even admitting that it does, it still leaves the circle described by the pole of the heavens undefined; for the instant we admit a variation in the obliquity, we have displaced the pole of the ecliptic from the centre of the circle described by the earth's axis, and unless we assume that the pole of the ecliptic again becomes the centre, again moves, and again strikes a fresh centre, we leave the problem as to the polar movement as undefined as ever. Thus, even granting the change in the plane of the ecliptic to really occur, we have not avoided the necessity of solving the problem relative to the course traced by the earth's axis.

But you have stated that the change in the latitude of stars near the ecliptic *proves* that the plane of the ecliptic does vary in such a manner as to account for the change in the obliquity.

This statement, I am aware, has been made and repeated in several astronomical works. It might be interesting if we could test how many of the writers of this statement have investigated its accuracy, and how many have merely copied it from one another.

Now the repetition by copying, of any statement which is not based on facts, may tend to promulgate error, but cannot be considered to give additional evidence, and we will now inquire as to the facts on which this assertion is based.

If the plane of the ecliptic is coming by a movement of this plane nearer into coincidence with the plane of the equinoctial, it follows that the north pole of the ecliptic must be moving somewhere in the direction of 6 hours R.A. This movement of the pole must cause all stars on the meridian of 6 hours R.A. to increase their north latitudes about 45" per century, whilst those stars having 18 hours R.A. will decrease their north latitudes in like manner. Thus this slow movement of the pole of the ecliptic would be very similar in its effects on stellar latitudes to the change produced on stellar declinations by a polar movement of 45". Thus stars having 0 and 12 hours H.A. would not have their latitudes changed by a movement of the pole of the ecliptic, which would bring it nearer the pole of the heavens.

These conclusions being correct, we adopted the following system of investigation in order to find the evidence relative to changes in latitudes:

We took at random four or five stars, having a longitude nearly the same and about 90°, and we compared the changes in latitude shown by these stars between the dates of Ptolemy's and Ulugh Beigh's catalogues. Between the catalogues of Ulugh Beigh and Hevelius, and by converting declinations obtained by latest observations into latitudes, we obtained another comparison. We then examined the changes in latitudes thus obtained with the changes exhibited by a similar comparison on stars differing 180° in longitude from those previously examined. This process was then adopted on stars differing about 5° in longitude from the former series, so that a very extensive test was obtained. We next selected nearly all the stars of Ursa Minor, Draco, Cepheus, and others near the pole of the ecliptic, and endeavoured to trace by changes in latitudes what had been the movement of the pole, and hence the plane of the ecliptic. After many months' careful research we found that there was no evidence worthy of the name that could be made to prove from the change in stellar latitudes that the plane of the ecliptic has varied during the past 1700 years. Other evidence of this variation there may be, but it is not from changes in latitudes of stars.

The pole star is so remarkable a star, and is so situated as to be at all times visible in northern latitudes, that it must have attracted attention even in ancient time. Towards this star the pole of the ecliptic ought to have moved, since the days of Ptolemy, about 20'. Thus in Ptolemy's Catalogue it ought to have 20' less latitude than at present; yet what are the facts? Tycho Brahé gives 66° 2' for the latitude in 1600, Hevelius 66° 3' for 1690; therefore Ptolemy's latitude ought to have been 65° 43', instead of which it is 66°.

Again, comparing many of the other principal stars, we obtain similar results, even when modern catalogues are compared; but this is no new statement, for we have merely to examine the *Philosophical Transactions*, vol. xxx. page 329, to find that Edmund Halley, than whom no more careful astronomer existed, remarked this fact and brought it before the notice of geometricians.

Another objector made this similar remark:

'That the plane of the ecliptic does vary is well known, not only by the changes in stellar latitudes, but by the change in the longitude of the ascending nodes of the various planets. When, then, you ascribe the variation of the obliquity to a movement of a pole of the heavens, you are opposing yourself to well-known facts.'

Our reply was:

'I do not deny that the plane of the ecliptic may vary, but I do deny that it is *proved* to vary by the change in the latitude of stars. We have four catalogues of stars extending over about 1500 years, in which the latitude of stars are given—viz. Ptolemy, Ulugh Beigh, Tycho Brahe, and Hevelius—and the comparison of latitudes in these is therefore easy. Also we can convert modern declinations into latitudes, and compare these results with either ancient or modern observations.

'If it were true that the plane of the ecliptic did vary its position 45" per century (still you must remember that the polar circle would remain without a centre), we ought to be able to trace exactly the course of the pole of the ecliptic by the changes in co-latitudes of the stars.

'We ought to find that since the time of Ptolemy the pole of the ecliptic ought to have moved away from stars having about 270° of longitude, and towards stars having 90° of longitude. It ought not to have changed its angular distance much from stars having 0° or 180° longitude: and this change ought to be clearly indicated, the plus and minus, on meridians differing 180°, being clearly marked.

'There is no mistake about the direction in which the pole of the heavens is moving when we *approximate* only to results. It is evident, from the decrease in north polar distance of all stars near 0 hours right ascension, and an almost corresponding increase in north polar distance of stars having 180° right ascension, whilst stars having 6 or 18 hours right ascension do not vary their north polar distance, that the pole is moving somewhere towards the meridian of 0 hours right ascension.

'Can you give me any evidence of the same kind relative to the pole of the ecliptic, or have you merely copied the statement that the changes of latitude prove that the plane of the ecliptic does vary?

'If you select two or three stars only, and draw a

conclusion from their changes in latitude, you may claim to prove any movement of the ecliptic; but for a change to be *demonstrated*, we must not *select* our evidence, but should examine a multitude of stellar changes of latitude.

'Thus you might select two such stars as Castor and Pollux, and claim to prove that the pole of the ecliptic had moved towards them, because the latitude of these stars is greater in the catalogue of Hevelius than in that of Ptolemy—Castor by Hevelius being given 10° 4' 23" north latitude, by Ptolemy 9° 30'; whilst Pollux is by Hevelius given 6° 40' 29", by Ptolemy 6° 15'. From the changes in latitude of these two stars, it would appear that the pole of the ecliptic had moved towards them 34' or 25'; yet during the same interval it has moved towards Polaris only 3', although Polaris is on nearly the same meridian of longitude.

'If, however, the pole of the ecliptic be moving towards that part of the heavens in which the stars Castor and Pollux are situated, it must be moving away from that part in which such stars as Vega and α Aquilæ are situated. This, you will urge, is a fact; for Vega has more north latitude in Ptolemy's catalogue than it has in that of Hevelius, and so also has α Aquilæ; and possibly you may now consider that it is proved that the pole, and hence the plane of the ecliptic, has moved downwards towards about 90° longitude, and thus explains the change in the obliquity.

'Let us, however, examine other and even more

remarkable stars. We find, from comparing the latitudes of the stars of Ursa Major as given by Ptolemy and Hevelius, that α Ursæ Majoris has 40′, β 37′, γ 38′, δ 38′ more north latitude by Hevelius than these stars are given by Ptolemy. Hence it appears that the pole of the ecliptic has moved towards these stars during the interval nearly 40′, and as these stars have about 120° of longitude, this movement of the plane of the ecliptic appears proved; and we might suppose that we could now trace the direction of the movement of the pole of the ecliptic by these changes in latitude. But if the pole has moved 40′ towards α and β Ursæ Majoris, it must also have moved about 40′ away from the stars of Cygnus, which are on the opposite side of the pole of the ecliptic; yet we find that it has moved only 2½′ from α Cygni, and 4′ 52″ from β Cygni; whilst it has, if the above were true, moved 40′ in the opposite direction.

'But, strange to say, whilst by these changes of latitudes the pole has moved 34′ towards a meridian of about 100° longitude, 40′ towards one of 120°, only about 4′ away from one of 290°, it has, as shown by the stars of Taurus, having also about 90° longitude, moved away from this part of the heavens; for α Tauri, having south latitude, indicates by its change of latitude that the pole has moved away from it 19′ between those dates at which it was supposed to move towards it, whilst γ Tauri shows the same position exactly at the time of Hevelius that it did in the time of Ptolemy. Nor is this a solitary instance; for Sirius, probably the most noticeable star in the heavens, and one to which ancient astronomers devoted great attention, indicates that the pole of the ecliptic has moved away from it no less than 20′ since the time of Ptolemy. β Canis Majoris shows a movement in the same direction of 4′, γ Canis Majoris a movement in the same direction of 14′.

'You cannot adopt a more severe test as regards the proof of there being or not being a change in star latitudes, that proves the plane of the ecliptic to be varying in one particular manner, than that which we adopted of selecting all the important stars within in 40° of the pole of the ecliptic, and noting the changes in latitude of these, checking each star by another differing 180° in longitude from the former. Thus, for example, take η Ursæ Majoris in Ptolemy's catalogue; it is there given a longitude of 29° 50′ of Leo = 149° 50′, and a latitude of 54°. By Hevelius and also by Tycho this same star is given 54° 25′ latitude; consequently the pole of the ecliptic ought to have moved towards this star in the interval between 140 A.D. and 1600 A.D. In the almost opposite direction, however, is α Cygni, which has in Ptolemy's catalogue a longitude of 9° 10′ of Aquarius = 309° 10′, and a latitude of 60°; consequently if the pole of the ecliptic moved towards η Ursæ Majoris 25′, it ought to have moved away from α Cygni 25′, and the latitude of this star ought to have been found by Tycho and Hevelius 59° 35′. Instead of which, both Tycho and Hevelius give 59° 57′ for the latitude, and thus indicate that the pole has moved away from this star 3' only.

'What conclusion are we to come to when we examine the stars of Cassiopeia? These stars have in Ptolemy's catalogue a longitude of only a few degrees of Aries; consequently their change in latitude ought to be very little, if any, just as the change of north polar distance is slight in those stars having 6 h. and 18 h. right ascension; but what are the facts?

'The star α in this constellation has decreased its latitude 8', β 27', γ 13' between the dates of Ptolemy and Hevelius; thus indicating that the pole of the ecliptic has moved towards these stars at the same time that it moved nearly in the opposite direction. Where, then, is your proof?

'Perhaps you will say from more modern observations. But comparisons made between two catalogues separated by an interval of about 1500 years ought to give us proof if it existed; and if you make use of the formula for converting declinations into latitudes, you may convert modern declinations into latitudes, and test in exactly the same manner as before, and, as far as I have gone, with the same results.

'It has been stated that stars should be selected near the ecliptic in order to test this change in star latitudes, but I think it will be granted that stars near the *pole* of the ecliptic ought to give more distinctly the evidence of any change. Having devoted much time to the investigation of these changes in latitude supposed to prove the change in the plane of the ecliptic, I send you a list of nineteen stars, all within 15° of the pole of the ecliptic, and around it in every direction.

Stars of Draco.	Ptolemy.		Ulugh Beigh.		Tycho.		Hevelius.			Longitude 1660.	
α.	76°	30′	76°	13′	76°	17′	76°	17′	17″	229°	46′
ν.	78	30	78	21	78	15½	78	14	20	245	52
β.	75	40	75	30	75	21	75	20	15	247	7
ξ.	80	20	80	0	80	21½	80	21	30	260	4
ι.	64	40	65	21	65	18	65	21	5	130	0
γ.	75	30	75	0	75	3½	74	59	37	203	0
σ.	81	40	81	45	81	53	81	48	12	359	24
δ.	83	0	83	0	82	49	82	44	15	13	54
ε.	78	50	79	0	79	51	79	27	50	28	18
ρ.	77	50	77	36	78	9½	78	9	0	15	48
θ.	80	80	80	30	80	54	80	57	12	27	10
π.	85	30	66	27	66	36	66	21	20	152	0
ψ.	84	30	84	12	83	4½	84	6	8	04	9
χ.	83	30	83	24	83	2½	83	29	40	71	45
φ.	84	50	84	42	84	48	84	48	20	60	45
α Ursæ Minoris (Polaris)	66	0	66	27	66	2	66	3	0	83	51
η.	78	0	78	57	78	32	78	29	0	180	15
θ. *	74	40	74	30	74	11½	74	29	0	180	47
υ.	81	20°	82	0	83	5	83	11	0	46	13
τ.	80	15	80	15	80	58	81	2	0	48	22

'If any star indicates by its change of latitude that the pole of the ecliptic has moved towards it, of course a star on the opposite side of the pole ought to show that the pole has moved away from it an equal amount. The stars given are those in the catalogues of Ptolemy, Ulugh Beigh, Tycho Brahé, and Hevelius, and cover a space of above

* True latitude above 89° (see Bailey's Catalogue).

1500 years; and during this interval of time there is not any evidence from these stars of such a change in the position of the ecliptic as has been assumed.

'Here is a list of the stars, with their latitudes, as given by the various observers. Examine them yourself and search for the evidence; but you must not forget that even granting a variation in the plane of the ecliptic, still this will not alter the truth of the polar curve.

'Probably you may say that the proper motion of the stars will explain these discrepancies; but if we assume that the want of accordance is due to this cause, we cannot claim by other stars to prove that there is a change in the plane of the ecliptic.

'I may now mention that these errors in star positions, and also a very great number of the so-called "proper motion" of the fixed stars, are due to the same cause as that which accounts for the glacial epoch of geology, viz. the movement of the earth's axis in a manner that has not hitherto been correctly interpreted; and as this subject is solely a geometrical one, it admits of proof as clear as that the pole of the ecliptic is not the centre of polar motion.*

'The change in the longitude of the ascending node is also affected by the same cause, as are many other important problems. Until, however, it is clearly understood that the pole of the ecliptic is not the centre of polar motion, and that the pole traces a

* Will be demonstrated in our next work, *The Cause of the so-called Proper Motion of the Fixed Stars.*

circle larger than that formerly assumed, other and more complicated problems depending thereon are not likely to receive much attention.

'Thus we reply that it is probable that the plane of the ecliptic does vary, but that changes in star latitudes give proof that it does, is not correct; for even granting this plane does move, still it leaves the course traced by the pole as obscure as ever, unless we define the radius of the circle which it is supposed to trace.

'You are of course aware that the precession of the equinoxes *ought* to produce a uniform increase in the longitude of all stars, and it has been supposed that the rate of about 50" annually gives us a close approximation to the true value. We have pointed out that the pole of the ecliptic is not the true centre of polar motion, and that another point 6° from it is the true centre. Naturally, then, we should obtain some strange inconsistencies when we came to examine the longitude of stars near the supposed centre of polar motion, and expected to find a uniform change in longitude. Take, for example, the longitudes of the stars of Draco, as given by Ptolemy and Hevelius or Tycho, and we have α Draconis 131° 10' and 152° 32'; difference, 21° 22'. β Draconis 203° 10' and 247° 7'; difference, 23° 57'. γ Draconis 239° 40' and 263° 9'; difference, 23° 29'. δ Draconis 350° 20' and 13° 25'; difference, 23° 5'. ω Draconis 111° 40' and 130° 22'; difference, 18 °42', this last star being near the assumed centre of polar motion. Is there uniformity of change here?'

The same writer forwarded the following reply:

'It seems true, as you point out, that there are very great inconsistencies in the longitude of stars near the pole of the ecliptic, for there is as much as 5° difference in the rate of some of the stars you have mentioned; but I do not see how the course which the pole traces has anything to do with star longitudes, as they are all counted from the first point of Aries.'

The following somewhat lengthy demonstration was sent in reply:

The actual value of the precession is due to the amount and direction of the polar motion and to the value of the obliquity, but it is independent of the position of the centre of polar motion.

To demonstrate this problem you may proceed as follows:

Let E be the pole of the ecliptic, P the position of the pole of the heavens at any date, P' the position of the pole of the heavens one year afterwards, consequently P P' the amount of polar motion in one year. P E would be a portion of the solsticial colure when the pole was at P, P' E a portion of the solsticial colure when the pole was at P'. Hence the angle P' E P will represent the advance of the solsticial colure, due to the polar motion from P to P'; and as the line of equinoxes is at right angles to the solsticial colure, it follows that the angle P' E P represents the precession of the equinoxes due to a polar movement from P to P'.

Thus the angle P' E P is the item to obtain, and it is dependent alone on the values of P P', P E, and P 'E, but is of the same value, no matter whether E, A, or B is the centre of the circle of which P P' is the arc.

The effect, however, of the precession of the equinox has been described to be the same as if the sphere of the heavens rotated once from right to left in 25,860 years round the poles of the ecliptic.

This description, however, would be correct on one condition only, viz. that the pole of the ecliptic was the centre of polar motion ; but if it be not the centre, then the effect might be *nearly* the same as though the whole heavens rotated once from right to left round two points which are 180° apart, and are not coincident with the poles of the ecliptic.

What are some of the differences which would result may now be demonstrated.

In the following diagram we take E as the pole of the ecliptic distant 23° 28' from P the pole of the heavens ; C T L I the ecliptic itself, all parts of which are 90° from E. Taking the apparent rotation of the sphere of the heavens round E as a centre, we should have for a small portion of this movement P moved to P', A to A', C to C', T to T', L to L', &c.

If the arc P P' were 20", then would T T' be 50"·1 = L L' = I' I = C C'. That is, all stars on the ecliptic would appear to be carried over exactly equal arcs, because all parts of the ecliptic are equally distant from the pole or centre of rotation.

Thus if the pole of the ecliptic were the centre of

polar motion, all stars would be found to vary their longitudes uniformly, and the effects of the precession

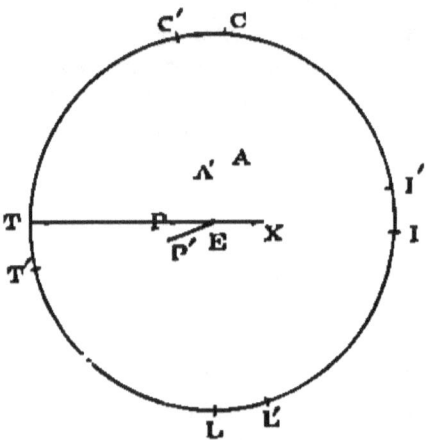

would be correctly described as the same in effect as though the sphere of the heavens rotated slowly round the poles of the ecliptic.

When, however, we know that the pole of the ecliptic is not the centre of polar motion, we can at once perceive how great a confusion must result if we assume that it is so, and when we find that a point x, 6° from the pole of the ecliptic, corresponds to the centre, we can demonstrate what results must occur.

Thus, for example, the points c and L on the ecliptic would be 90° from x if E x c were a right angle; but the points T and I would be 96° and 84° from x; consequently a star at c or at L would appear to be carried over a greater arc by the supposed rotation of the sphere than would a star at T or I. For example, if a star at c were carried over an arc of 50″, a star at T would, though also on the ecliptic, be carried over an arc of only 49″·86. It is perfectly true that a meridian of longitude passing from E, the pole of the ecliptic, to that part of the ecliptic which is for the time being the initial point for longitudes and right ascensions, will, as it shifts with this point, cause a uniform increase in the longitudes of every star in the heavens.

Thus in the annexed diagram, suppose E the pole of the ecliptic, $a\,b\,c\,d$ the plane of the ecliptic, and P the pole of the heavens.

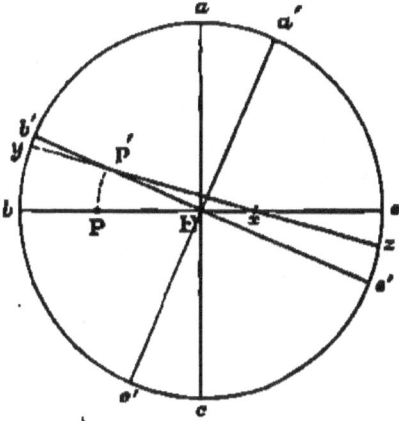

E a will be the initial meridian, from which star longitudes are counted.

Let us suppose the pole P to have moved to P′ by

the conical movement of the earth's axis; a *celestial* meridian will now pass through b' P' E e', and the initial meridian for longitudes will be E a', thus causing all stars to have increased their longitudes by the angle a E a'.

But this movement is *in* the earth, not in the heavens; and the fact of the polar movement being traced round a centre not coincident with E will cause the meridian which was at b P E e to occupy the position shown by y P' x z, where x is the true centre of polar motion, and not E; that is, this movement of precession acting alone causes this divergence of the meridian b P E e to y P' x z, and does not cause this meridian to occupy the position b' P' E e'.

The difference between the two new positions of the meridian b P E e, according as E or x is the centre of polar motion, causes great confusion in longitudes and right ascensions; for the meridian b P E e, to pass again through E when the pole has been carried to P', requires that the diurnal rotation should correct this difference,

The diurnal rotation, being in the direction opposite to that in which the meridian has been carried by the polar motion, will cause the meridian y P' x z to pass across E; or in other words, E will transit this meridian *before* it would if E were the centre of polar motion, because the meridian y P' x z is, when referred to diurnal rotation, in advance of the meridian b' P' E e'.

Now, as in the present day right ascensions are

determined entirely by meridian transits, and longitudes depend on right ascensions, you will perceive that there are some most important points to be considered in connection with this problem.

It is scarcely necessary to point out to you how many beautiful geometrical problems result from this investigation, and how interesting they become when we compare the results obtained from a knowledge of the position of the centre of polar motion, with the volumes of laborious and expensive tables of stars' proper motions in right ascensions, accumulated with so much labour by the recorders of facts.

When you read carefully this demonstration, I think you must perceive, that though the actual precession of the equinoctial point would be the same for a polar movement of 20″, no matter whether the pole of the ecliptic or the point x were the centre, yet very different results would occur as regards the changes in stellar longitude. And when we find that between Ptolemy's time and that of Hevelius we obtain a difference of 5° in what ought to be uniform, I must confess I see no cause to alter the conclusions to which I had previously arrived.

Probably no part of this investigation has occupied a longer time than has the inquiry relative to the changes in stars' right ascensions and declinations; but as this is one of great importance to the practical astronomer, it is one requiring great care.

It will be evident that the great circle of the sphere which would indicate the greatest *apparent* motion of rotation would not be the ecliptic, but a great circle inclined to the plane of the ecliptic at an angle of 6°; this circle would at the present date intersect the plane of the ecliptic at about 356° and 176° right ascension.

If the pole of the ecliptic were the true centre of polar motion, the effects of the change in direction of the earth's axis would—as has been said—be the same as if the sphere of the heavens rotated, and thus we should have to consider which was the more probable, that this movement was in the earth or in the sphere of the heavens. Now, it is a singular fact that we are able to demonstrate that it is a movement in the earth, and not in the heavens, and this proof is a geometrical one, and may be given thus.

If the pole of the ecliptic were the centre of polar motion, the effects would be the same as though the whole heavens rotated, and a star at the pole of the ecliptic would always remain fixed, whilst all others rotated round it in 25,860 years. With a point 6° from the pole of the ecliptic as the centre of polar motion, the effects if the heavens rotated would be to carry a star which was coincident with the pole of the ecliptic away from that point, and this star would be carried annually over an arc of 4″ if the polar movement were 20″. Thus, as since the earliest records the pole has moved 10°, we should find a star at the pole of the ecliptic would have moved 2° by this rotation of the heavens. But as the pole of the ecliptic will not alter its position unless the plane of the ecliptic moves, and as a star would not alter its position from a fixed point unless the sphere of the heavens moved, it follows that the rotation of the sphere of the heavens cannot be the cause of the precession; for if it were, stars would move towards and away from the pole of the ecliptic; therefore we can explain this movement only by assigning to the earth's axis a change in direction.

It is not correct to state that the effects of the precession are the same as if the whole heavens slowly rotated round the poles of the ecliptic, nor are they the same as if the heavens rotated round the true centre of polar motion. The results that are different produce those variations in longitudes when we compare olden catalogues, and those variations in right ascensions when we compare modern catalogues which hitherto have been generally attributed either to the errors of olden astronomers, or to a proper motion in nearly every one of the fixed stars; and this confusion will continue until the true motion of the pole is known, and the true centre of polar motion correctly localised.

CHAPTER VIII.

OBJECTIONS RAISED AND CONSIDERED (CONTINUED).

Having requested a distinguished astronomer to favour us with his explanation of the exact course traced by the pole of the heavens, he, with that courtesy which all inquirers have ever met at his hands, informed us, 'That the direction of the movement of the terrestrial pole in any one year on the surface of the globe is at right angles to the great circle connecting the plane of the terrestrial pole with the plane of the ecliptic pole in that year.'

Having pointed out that this definition fixed the centre of the circle traced by the pole on a great circle joining the two poles at that time, and having then requested to know where the centre of this circle was supposed or assumed to be localised, we pointed out that the pole of the ecliptic could not be the centre of the circle traced by the earth's axis, in consequence of the decrease of the obliquity, and we then questioned whether the true course of the pole during its whole revolution was really understood.

We did not then consider it necessary to give either the formula which proved what the course of the pole really was, nor to refer to anything more than the incompleteness of the definition relative to a whole revolution of the pole.

In reply to this inquiry and statement, the same writer gave the following answer: 'The motion of the earth's pole is perfectly understood. For a short time, as a few years, it may be considered as a circle, but for a long time it is not a circle.'

We regret to be compelled to differ from this distinguished authority; for, in spite of the statement made in his answer, we think this reply clearly indicates that the course of the pole for a long time is not understood, nor can it be understood whilst the curve is an undefined curve. In order, however, to definitely answer those persons who assured us that the movement of the pole during its entire course was thoroughly understood, we applied to the first astronomical authority to know if he would inform us what would be the obliquity at the date 10,000 A.D., and what it was 14,000 B.C.

The reply was definite: 'I cannot undertake to say.'

If we know the curve traced by the pole during its whole course, we *can* state what the obliquity would be at the date 14,000 B.C. within a very few minutes. When, however, this curve is undefined, it is impossible to say.

Among the various writers who favoured us with their opinions or objections, was one who was under the impression that he had discovered an error of a very palpable kind. Having found that from our data we gave 45" as the amount of decrease in the obliquity during *the present* century, he at once assumed that this rate would be uniform, and instead of examining either the diagram we forwarded, or the formula already given, he proceeded to multiply 45" by 140, and then pointed out that instead of obtaining 12° variation in the obliquity, we could obtain only 1° 45' for 14,000 years.

Such an error or absence of observation would be merely amusing, if it were not that the writer, from his position, had the opportunity of misleading many even less well-informed than himself, and thus misrepresenting a problem which probably might have been examined by some person more competent to judge than was this writer.

More than one of our correspondents have urged the following objection:

'It is well known that the revolution of the equinoxes occurs in exactly 25,868* years, whilst you state 31,000 as the time. How could such exactitude be obtained, if the real course of the pole were not known?'

In answer to this objection we wrote:

'The period 25,868 years is obtained by a rule-of-three calculation. You will note in the article you quote that the annual precession is supposed to be 50"·10 per annum. If you work out this sum, you will find that if in one year 50"·10 are passed over, then

* Herschel's *Outlines*, Art. 316.

it will occupy 25,868 years to complete 360°. This is not, as you seem to consider, an abstruse or difficult problem based on long and careful observation, but a mere proportion, based on the assumption that 50″·1 is the annual rate of the precession, and that it is and always has been uniform; which, however, it is not.

'There is an error in the present method of calculating the precession.

'The value of the precession of the equinoxes is due entirely and solely to three causes:
1. To the amount of polar motion.
2. To the obliquity of the ecliptic.
3. To the direction of polar motion.

In consequence of its having been assumed that the pole of the ecliptic is the centre of polar motion, and that consequently the pole of the heavens never varies its distance from this pole, it has been concluded that the direction of polar motion is exactly at right angles to the arc joining the two poles, and the value of the precession is calculated on this assumption.'

Another correspondent favoured us with the objection, that if this new movement of the pole were accepted, it would interfere with many theories.

In reply we stated that the movement which we had submitted was based on observations which were unquestioned; that the investigation was a rigidly geometrical one, and was neither a theory nor based upon a theory; and as geometry is the most exact of sciences, it was scarcely a sound objection that probably some theory might be interfered with.

VALUE OF THE OBLIQUITY.

An able geometrician requested us to explain a discrepancy between the formula already given, and by which we found the value of the obliquity, and certain recorded observations. His remarks were as follows:

'Upon working out by your formula the value of the obliquity, I must own that, as you say, I find the results correct to a fraction of a second for the present date, 1870, and correct for many years in the past. When, however, I compare the results obtained by your formula with Bradley's observations in 1755, I find a discrepancy of about 4″; Bradley giving 23° 28′ 15″·3, and your results giving about 3″·5 more than this for the obliquity. Do you think Bradley could have made an error of 3″·5 in his observations?

'Also I find that it is stated in various works that Ptolemy found the obliquity at 140 A.D. 23° 48′ 45″, whereas by your formula it ought to have been greater by some minutes. Now as this observation of Ptolemy's is the only one we have from that remote date, and as it is given to seconds, I place great value on it, and think the discrepancy between it and your results most important.'

Our replies to these questions were as follows:

'First, as regards Bradley's observations, I do not think 3″ or even 4″ a very large error for Bradley to have made in the obliquity, considering the imperfection of his instruments and the uncertainty of refraction; for I find the following are given as the obliquity found by other observers about the same date:

By the French astronomers for 1755	. . .	23° 27′ 19″
By La Caille	„ 1748 . . .	23 28 20
By Godin	„ 1734 . . .	23 28 20
By Cassini de Thury	„ 1743 . . .	23 28 26

'It is evident that these differences are either due to errors of observation or to errors in the tables used for corrections. On examining Bradley's observations, I find he took the latitude of Greenwich Observatory to be 51° 28′ 39″·6; consequently the altitude of the equinoctial on his meridian would be 38° 31′ 20″·4. If the obliquity were 23° 28′ 15″·3, the sun's greatest meridian altitude during the year would be 61° 59′ 35″·7, and the least would be 15° 3′ 5″·1. Now as the refraction for 61° 59′ 35″·7 is about 8″·2, and for 15° 3′ 5″·1 about 3′ 34″, it follows that the *apparent* greatest and least meridian altitudes of the sun would have been 61° 59′ 43″·9 and 15° 6′ 39″·1, without allowing for parallax. In order, then, to find whether Bradley's *observations* were in error, we must find what refractions he used. Fortunately, I possess a work on astronomy, published in 1744, in which it is stated that the following were the refractions used for an altitude of 15° by the following astronomers:

La Caille	. .	3′ 49″
Bradley	. .	3 30
Halley	. .	3 17
Newton	. .	3 16
Flamstead	. .	3 0
Taylor	. .	3 20

In the table of refractions given in the *Fundamenta Astronomiæ*, we find 3′ 31″·4 as the value of refraction used by Bradley for altitudes of 15°, and as this value is about 3″ less than it ought to be, the error would explain the difference between Bradley's recorded observation and the results obtained by our formula, and would cause the two to agree to a second. Thus we are enabled to tell almost to a certainty, even one hundred years ago, what has been the error in a table of refractions used by former astronomers.

'Next, as regards Ptolemy's recorded obliquity.

'I fear that you have adopted the common error of copying from some popular work the statement that Ptolemy found the obliquity exactly 23° 48′ 45″.

'If you examine a copy of Ptolemy's work (*Almagest*) in the British Museum Library, you will see that he states he found the difference in altitude between the summer and winter of the sun 47°⅔ to 47°¾, which would give 23° 50′ or 23° 52′ 30″; and as Ptolemy knew nothing of refraction, and was unable to read an angle on his instruments nearer[*] than 10′, it is not correct to state that he found the value of the obliquity to within a second.

'You are incorrect when you state that Ptolemy's is the only ancient record of the obliquity. If you examine the back numbers of the *Philosophical Transactions*, you will find a multitude of records from

[*] See Preface to Ptolemy's Catalogue, *Memoirs of the Royal Astronomical Society*, vol. xliii. p. 17.

ancient times. From among these I give you the following:

	Arths of the circle =	24°	0′	0″
Strabo, A.D. 30.				
Geminus do.	,, ,,	24	0	0
Tatius do.	,, ,,	24	0	0
Proclus do.	,, ,,	24	0	0
Abraham Abnezda, A.D. 30	,, —	21	0	0
Pappus, A.D. 390	,, ,,	23	50	0

and many others, all giving about 24° for the value of the obliquity about A.D. 30.

'Why Ptolemy's observation alone should have been quoted, whilst many other observations of the same date appear to be disregarded, I cannot tell. Such omission, however, has led you into an error, and may probably mislead others.

'If you work out the obliquity by my geometrical formula, based entirely on the movement of the pole in a circle, the radius of which is 29° 25′ 47″, you will obtain for the date 1437 A.D. an obliquity of about 23° 32′ 13″, and you will find that Ulugh Beigh gives 23° 31′ 58″ for his observations at that date, giving a difference only of 15″; a difference easily explained, either by there being no correction then made for refraction, or on account of imperfections in instruments.

'Thus there is not a discrepancy such as you suppose between any *dependable* observations and the formula I have given; and when you consider that I make no use of any recorded observations in order to obtain the results, but start from a condition no less than four hundred and twenty-seven years in advance, and yet obtain by calculations based on purely geometrical data results which agree to the tenth of a second with the observations of the past thirty years, you must grant that I have some grounds for placing confidence in the accuracy of these conclusions.'

Another correspondent offered the following objection:

'In consequence of the difference between the tropical and sidereal year being dependent upon the value of the annual precession, it ought to follow that as there would be a slight difference in the resulting value of the precession by the polar movement which you advocate from that resulting from the pole moving in a circle round the pole of the ecliptic as a centre, so ought there to be found some error in the value of the equinoctial year when long periods, such as a century, were referred to. Now as the year is known to be exactly 365 days 5 h. 48 m. 42·7 s. in length, and it is also known that during the past century it has not varied one-tenth of a second, it follows that the present polar movement must be correct.'

Our reply was as follows:

'In answer to your statement that during the past century every item connected with the length of the tropical year has been found correct, I must refer you to the note on the last page of Herschel's *Outlines*, on the subject of the tropical year. This note you will find as follows:

"These numbers differ from those in the *Nautical*

Almanac, and would require to be substituted for them to carry out the idea of equinoctial time as above laid down. In the years 1826-1833 the late eminent editor of that work used an equinox slightly differing from that of Delambre, which accounts for the difference in those years. In 1834 it would appear that a deviation, both from the principle of the text and from the previous practice of that ephemeris, took place, in deriving the fraction of 1834 from that for 1833, which has been ever since perpetuated. It consisted in rejecting the mean longitude of Delambre's tables, and adopting Bessel's corrections of that element. The effect of this alteration was to insert 3 m. 3·68 s. *of purely imaginary time* between the end of the equinoctial year 1833 and the beginning of 1834; or in other words, to make the interval between the noons of March 22 and 23, 1834, 24 h. 3 m. 3·68s. when reckoned by equinoctial time. In 1835, and in all subsequent years, a further departure from the principle of the text took place, by substituting Bessel's tropical year of 365·2422175 for Delambre's. Thus the whole subject has fallen into confusion, and we have to choose between resorting to the original design in its integrity, or to continue the present practice (eschewing all further change) for future years."

'When you become acquainted with these facts, I believe you will consider it necessary to modify your assertion that the tropical year has been correctly known to one-tenth of a second for at least a century;

also I believe you will perceive that your assumed facts are not correct, and therefore not only is your objection unsound, but I may state that, having devoted nearly a year to the one subject of Time, as affected by this polar movement, I find that the confusion here referred to by Sir J. Herschel is almost entirely caused by the assumption that the pole of the ecliptic is the true centre of polar motion, and to correct these errors has been one object of my investigation.'*

More than one writer raised the following question:

'You have stated that unless the true polar motion were known, it would be impossible to predict the true value of the obliquity for a few years in advance; how is it, then, that in a work I possess it is predicted from 1800 to 1900 to within the one-hundredth of a second?'

"The work to which you refer is probably Baily's *Astronomical Tables*, page 165, where the mean obliquity is given for every year from 1800 to 1900. If you will examine Bradley's catalogue, you will find that Bessel, in his preface, has given the obliquity found by several observers (page 61), and he deduces for the annual rate $0''{\cdot}457$; therefore $45''{\cdot}7$

---

* Every competent geometrician will perceive that the true value of a sidereal day can only be known when the true centre of polar motion is known. As the pole of the ecliptic has hitherto been erroneously assumed the centre, the successive transits of the pole of the ecliptic have been supposed to give a uniform measure of time; this assumption is incorrect.

for a century for the decrease. He also deduces 23° 27′ 54″·78 for the obliquity at 1800. No very great skill is required to subtract 45″·7 from the obliquity given for 1800, and thus obtain 23° 27′ 9″·08, which is given as the obliquity for 1900 in the table you quote; but, even now, this table disagrees with observation.'

Several other similar objections were urged by those who were unacquainted with geometry, and who even went so far as to state that they saw nothing erroneous in the two statements, 'that the pole of the heavens traced a circle around the pole of the ecliptic as a centre, whilst there was a constant decrease in the obliquity;' but these objections are of such a nature as not to require any reply.

About two years ago we had the honour of reading a paper on the geological effects of this movement of the pole before the Geological Society, and had the opportunity of hearing the opinion of geologists as regards the results. Unfortunately geologists, as a rule, are neither geometricians nor astronomers, and so we could obtain no criticism on the astronomical portion of the paper; no objections, however, were urged by geologists against the facts then brought forward, and thus no answers are needed.

We had also the honour to read before the Royal Astronomical Society a paper on the effects produced by this movement of the pole on the measurements of time; but as the subject was one which did not appear to have occupied the attention of any of the members present on that evening, we obtained no criticism thereon.

That we should have been favoured with some objections by our fellow members of the Royal Astronomical Society, had a full abstract of our paper been given in the monthly notices, is most probable; but unfortunately the paper, which was the result of upwards of four years' research, was disposed of in the notices in some dozen lines, and as no opportunity was given us of seeing or correcting this notice previous to its circulation, it contained a clerical error which entirely misrepresented the problem we had brought forward.

As, however, any off-hand discussion on a question of geometrical fact is not likely to be of much advantage in deciding on the soundness or unsoundness of a problem, an evening meeting at which some half-dozen papers are hastily read and criticised, is scarcely giving sufficient time or thought to what may be of importance. Thus, probably, after all a book is the best form in which a subject can be brought forward, as it has a wider field of circulation than a paper read before a society. A problem of this kind is one more suited to the calm investigation and searching criticism of the studio, than it is to the verbal fault-finding, which too often is the result of so hasty an examination as could alone be given to an evening paper. Having, however, considered that it was a duty we owed to our fellow-members of the Royal Astronomical Society to read first before them a brief outline of this problem, we now consider it a duty we owe to the problem itself to make it more generally known, and invite more general criticism than it has yet received.

The following amusing incident we trust we shall be pardoned for publishing; it so aptly describes the manner in which difficult problems are sometimes 'shelved' or considered settled and disposed of, that it may serve a useful purpose.

Some years ago a personal friend, who was a fair mathematician, asked us if he might attend a series of lectures we were then giving on practical astronomy, as he wished to learn something of the subject. He attended the course, and worked with us, and acquired a fair knowledge of the transit instrument and could work out a lunar. He was, however, always somewhat obscure about the difference between a star's right ascension and its longitude. Some time after, he obtained an appointment which, as it was connected with astronomy, gave him a certain position.

About this time we asked some geological friends to examine the problem herein submitted, and they agreed to do so.

After some weeks' delay, we were informed by these gentlemen, that as far as the geological matters were concerned, the movement was just what was required; but having submitted the problem to a 'competent astronomer,' he had decided that it was quite opposed to what was known, and was therefore untenable.

We could obtain no details as to what the objections were, and so could not answer them.

Shortly after this decision our astronomical friend visited us, for the purpose of obtaining some explanations relative to the annual changes of stars' declinations. He casually mentioned that some time previously there had been submitted to him a problem relative to the movement of the pole, upon which he was requested to give a decision. His account of the problem was graphic.

'It was,' he said, 'with reference to the pole tracing a circle round some point in the heavens which was not the centre of some circle. I looked over it, but could make neither head nor tail of it; and as it is unlikely anything new can turn up now, I copied from a standard work the description that the pole of the ecliptic is the centre of polar motion, and this other movement therefore could not occur. I believe I obtained great credit for my quick explanation.'

The problem thus decided and *proved* to be 'unsound' was our own; and whilst we could congratulate our astronomical pupil on his rapid progress, we could scarcely place much value on the result of the decision to which the geological gentlemen had been led.

A correspondent favoured us with the following singular view of the problem:

'I grant,' he wrote, 'that your formula does give with wonderful accuracy the value of the obliquity for all time, as far as we have records; also it does seem to supply exactly those climatic conditions required to explain the boulders and erratic phenomena. But may not the pole of the ecliptic move just so as to account for the variation in the obliquity, in the same manner as would occur if your interpretation of the polar movement be correct?'

We replied:

'I can scarcely think it possible that so singular a coincidence, or rather series of coincidences, could occur in connection with the movement of the pole of the ecliptic as to cause it to approach the pole of the heavens in exactly the ratio found for the curve we have traced out. To agree with recorded observation, the pole of the ecliptic cannot move uniformly; and you forget that if the pole of the ecliptic is moving, you are bound to find a centre for the circle traced by the earth's axis. *I also maintain that the facts brought forward by geologists relative to the Glacial Epoch require from astronomy an explanation as much as does an eclipse of the moon.* To state our inability to account for the climate of the glacial epoch merely because, by theory, one mathematician lays down as finite 1° 21' for the variation of the plane of the ecliptic, and another says 4° 3', whilst the variation of this plane has but little to do with the course traced by the earth's axis, seems feeble in the extreme.

'Thus I need not go farther than to point out that there is at present positive evidence that the course of the pole is a circle such as I have described; and by this course both astronomical and geological facts are explained, whilst by the other neither receive any satisfactory solution.'

The following remarks and replies passed between us and one of our correspondents:

'It is of course evident that the pole of the ecliptic cannot be the centre of polar motion when we know that the obliquity is variable, and it is only a popular way of describing this motion to call the pole of the ecliptic the real centre; but the centre is so *very near* the pole of the ecliptic, that for all practical purposes it may be taken as coincident with it.'

To this singular assertion we replied:

'You state that the centre of the circle is *so near*. How near? Is it within 5' or 5°? Is it on the 270° meridian of longitude, or the 90°, or on the 180° meridian, or where? How many degrees is it from the pole of the heavens, and at what rate does the pole of the ecliptic vary its distance from it, or do the two vary their distance?'

The answer was:

'I cannot tell you exactly where the centre of polar motion is; but there is no doubt that it is very close to the pole of the ecliptic, probably within a few minutes; for if it were not, there must occur great changes in the obliquity, which are known never to have occurred, as was most ably demonstrated by M. Laplace.'

That singular confusion relative to a change in the plane of the ecliptic being the only possible cause of a change in the obliquity had so possessed this writer, that the mixture in his mind of the centre of polar motion with the variation in a plane, rendered it impossible for him to perceive that unless the centre of polar motion could be defined and localised, the problem relative to the variation in the obliquity could not be solved, no matter what 'authority' was quoted.

That it should be called a 'popular' way of describing this movement is giving a very mild term to the use that is made of it. 1st. On every celestial map the circle is drawn, and round it is written 'Circle described by the pole in 25,860 years.' 2dly. The fraction of the year is obtained from this supposed circle. 3dly. The mean sidereal day is deduced from the same item, and the ratio 1·00000011 to 1 is thereby obtained, and on this ratio all our astronomical calculations are based. If this is a mere popular description, what, then, is an exact one?

More than one person has quoted the theories that have been so long accepted relative to the joint action of the sun and moon on the protuberant mass at the earth's equator, and have pointed out how these theories prove that the earth's axis must maintain the same constant angle to the plane of the ecliptic, never varying this angle.

Many of these persons have concluded their objections by stating that it was 'impossible' that the earth's axis could vary the angle which it made with the plane of the ecliptic.

To such an argument we could offer but the old reply, that we did not state that it was possible, we merely state that *it occurs*. It may be impossible, by some theory, for the earth's axis to vary the angle it makes with the plane of the ecliptic, but nevertheless it is a fact that it does vary this angle, and has done so during 1800 years at least. Thus, when we have facts recorded during 1800 years which prove that the angle varies, we believe it will be admitted that it is so much the worse for the theory which states it cannot do so, not, as some individuals would prefer, so much the worse for the facts. Theories should be invented to account for facts, not to give a reason for denying them. We believe that a slight course of geometry will convince these objectors that a variation in the obliquity means the same thing as a variation in the angle which the earth's axis makes with the plane of the ecliptic, and that to deny this variation is, after the long recorded observations which we have given, somewhat useless.

Another correspondent stated that one of the most beautiful mathematical theories he knew was the explanation, by theory, of the cause of the earth's axis tracing a circle round the pole of the ecliptic as a centre, and that nothing would make him give up this theory.

We replied that we also considered this theory very clever, but the unfortunate part of it was, that the event did not occur which it was supposed to explain. The theory was supposed to explain why the earth's axis traced a circle round the pole of the ecliptic as a centre, and as unluckily it does not do so, the value of the theory is somewhat deteriorated, at least in the opinion of those who value facts.

But we remarked, that if this writer examined the works of astronomy written about twenty or thirty years ago, he would find that M. Laplace was supposed to have demonstrated that the plane of the ecliptic varied to the amount of only $1° 21'$. If any writer thirty years ago had expressed an opinion that the plane might possibly vary as much as $1° 26'$, even he would have been excommunicated as a scientific heretic. Yet suddenly M. Le Verrier thinks he demonstrates that this plane can vary not $1° 21'$ only, but actually $4° 3'$; and now, not to know this demonstration is considered a proof of ignorance of the latest discoveries. Now as M. Le Verrier does not possess any new fact unknown to M. Laplace, and yet demonstrates another result, so we have full confidence, that when this question of the centre of polar motion is looked into, and it is found that it is no longer tenable to assert that the pole of the ecliptic is the centre of polar motion, the very same theories which demonstrate so beautifully why the pole of the heavens moves round the pole of the ecliptic as a centre, will demonstrate with equal skill why it moves round a point as a centre $6°$ from the pole of the ecliptic.

We would suggest that, as there is so much more land in the northern than in the southern hemisphere, and such large lead mines in the north, the centre of gravity of the earth does not lie in the plane of the equator. Perhaps this fact, combined with one or two other items, will enable theorists to deal with the problem now presented to them, and by a very slight modification of their present processes, be able to show that it is impossible that the pole of the ecliptic can be the centre of polar motion; and in fact that theory demonstrates that the only possible place for the centre of polar motion is a point 6° from the pole of the ecliptic, and having a right ascension of about 264° in 1870. It will be a great triumph for theory if it can accomplish this, as we trust it will as soon as these facts are established.

It may interest some readers to know that the old proverb that 'history repeats itself' is shown to be true by many examples which we could quote in connection with the present investigation. We give below the remarks of one amiable gentleman whose criticism we invited in connection with the movement of the pole as herein demonstrated, and these remarks are so very similar to others received from various sources, that we seem when reading them to have gone back to those old days when Padua and Pisa were the centres of scientific dogmas.

This gentleman characterises our rigid geometrical demonstration of what the curve traced by the earth's axis really is as a 'vague belief.' He then describes the movement of the pole of the heavens round the pole of the ecliptic *as a centre* as 'a well-established fact.' But we must let him state his own case.

'I have well considered your vague belief about the movement of the pole in a large circle, and I must own that as I find that during two centuries at least the astronomers of all nations have agreed that it is a well-established fact, that the pole of the ecliptic is the true centre of the circle described by the earth's axis, and that the best authorities have stated that the pole of the heavens never varies its distance of 23° 28' from the pole of the ecliptic, I must declare that no proof which you can bring will ever convince me to the contrary. And as to geology having anything to do with astronomy, I consider the two sciences have not the least connection one with the other.'

When such remarks emanate from an ignorant and unknown man, they are at first amusing; when, however, they are written by one whose position and occupation enable him to promulgate such statements in the lecture hall, a feeling of melancholy steals over us as we reflect upon the condition of those who study at the feet of such a professor.

We are never likely to arrive at the elucidation of any truth, when for calm and continued thought we substitute a mere war of words; but there is a limit beyond which even philosophical patience ought not to be expected to retain its equanimity, and we must be therefore excused if to this perhaps unconscious perverter of a problem we wrote somewhat plainly.

We replied:

'What you term a "vague belief" is a conclusion arrived at by perhaps the most rigid and exact geometrical investigation that has ever been applied to an astronomical problem. What you term a "well-established fact," which for 200 years has been agreed to, is a geometrical impossibility which no amount of "authority" can make to be true, any more than "authority" could make the three angles of a plane triangle less or greater than 180°.

'If you could realise the possibility of the course of the pole having been agreed for 200 years to be such as I have shown it to be, I believe you would then consider it as ridiculous to say it was not so, as to doubt now that the earth is spherical or spheroidal in form.

'You state that no amount of evidence will convince you. After a certain point in a demonstration has been reached, it becomes absurd to deny the truth of this demonstration, and to adhere to some old theory which not only has never been demonstrated, but can be shown to be untrue. Where is the proof that the pole of the ecliptic is the centre of polar motion, or is even near to it? In fact, those who continue "objecting" and refusing to accept as true a problem which is fairly proved, occupy with regard to such problem exactly the same position which those types of persons do who are not yet convinced that the earth is not a flat surface, and who formerly refused to believe in the rotation of the earth. Thus whilst it is a duty to fairly criticise a novelty, you should beware lest this criticism degenerate into an obstinate adherence to an old theory, which geometry proves to be untenable. Such a proceeding, instead of gaining a man the reputation of a philosopher, will most probably cause him to be morally pinned up in a cabinet, and classed with those learned professors of old who, in spite of Galileo's telescope, yet struggled hard to prove that there were no satellites to Jupiter, and that Saturn's ring was a vague belief.

'If you deny that the course of the pole is such as I have described it, you must believe it to be some other curve—probably, as you state, a circle having for its centre the pole of the ecliptic.

'I have given you the proof and evidence why I believe the curve to be such as I have described it to be; will you favour me with the evidence on which you believe the pole of the ecliptic is the centre of the circle?

'That it has been *supposed* to be so for 200 years is scarcely a proof, considering that for 1800 years at least the earth was *supposed* to be stationary.'

Our correspondent did not reply, and so we were not favoured with his 'proofs.'

A distinguished astronomer, whose opinion is worthy of every consideration, having seen a small portion of this work, made the following remarks:

'You attribute the change of obliquity to a motion of the pole of the earth which is not recognised. But this is not the cause of the change of obliquity. That change arises from a motion of the pole of the ecliptic. The plane of the ecliptic, which is the plane of the earth's motion round the sun, is not invariable, it is disturbed by the attraction of the planets; and thus the celestial ecliptic and the pole of the ecliptic are changed; the latitude of stars are changed also by it.'

On examining these remarks, we cannot avoid perceiving that there is the same want of distinctness with regard to the course of the pole of the heavens. If, as this writer states, the motion of the pole of the earth does not produce a change in the obliquity, then of course the pole of the ecliptic must be the centre of the polar circle. The remark that the cause of the obliquity of the ecliptic decreasing is in consequence of this plane varying, is the idea which has so long been accepted, and which ignores the fact that the variation of this plane will cause no variation in the obliquity if the pole of the ecliptic be really the centre of polar motion, and if it be not the centre, the nature and character of the curve which the pole of the heavens traces is undefined.

Of course, if the plane of the ecliptic varied, the latitude of stars must vary; but that they do vary, so as to give a proof of any regular change in the plane of the ecliptic, we have already shown is not a fact.

The course which the pole of the heavens traces can no longer be considered a circle, the centre of which is the pole of the ecliptic. We have demonstrated what this curve is, when we show how the distances vary from the pole of the ecliptic. We may therefore reasonably ask the writer of the above remarks to give us some demonstration, at least to prove how it is that the motion of the pole of the heavens *does not* produce a change in the obliquity. We know this pole moves 20" annually over an arc, which is proved to be a portion of a circle the centre of which is 6° from the pole of the ecliptic; where, then, is the evidence on which we can assert that this movement causes no change in the obliquity?

Some of the correspondents, whose opinion we have requested relative to portions of this problem, appear to have entirely misunderstood our object in asking for their criticism. They appeared to be under the delusion that we wished them to examine our work, in order to note whether or not we had correctly copied the works of former astronomers, made some trifling alterations in the wording, and then brought out the book as a new one of our own.

When, then, they found that in this manuscript there was a statement relative to the course traced by the pole round the pole of the ecliptic, which would cause a considerable change of climate on earth, they did not examine the evidence on which we came to this conclusion; but they referred to the first astronomical book that came to hand, copied from that the statement 'that the pole of the ecliptic *was* the centre of polar motion, and never changed its distance of 23° 28' from it,' and informed us that all we had written seemed correct except that we had made a mistake relative to what the movement of the pole was, as we should see if we referred to So-and-so's Popular Astronomy.

We believe we need not point out that this work is not a copy. It is not a book written, as it is sometimes called, 'with a pair of scissors,' but in it there is a new problem, the truth or falsity of which is not to be decided upon according to the agreement or disagreement of this problem with what has hitherto been supposed, but it must be judged according to its geometrical accuracy, and according to its power of explaining facts.

It is quite similar to the practice of olden critics to decide that a novelty must be wrong, because it is different from what it may be the fashion to consider orthodox at the time; for science has its fashions, and geometry just now seems sadly unfashionable, considering what is accepted relative to the centre of polar motion. The rotation of the earth was objected to, because it did not agree with Ptolemy's theories; and so we have found the views here put forward have by some persons been supposed wrong, because they found in a printed book that something different had hitherto been stated to be the real thing.

CHAPTER IX.

CONSIDERATIONS RELATIVE TO GEOLOGICAL TIME AND THE ANTIQUITY OF MAN.

GEOLOGISTS have made many attempts to estimate the length of time that has elapsed between present dates and some of the later geological events. The date of the glacial epoch especially has occupied much attention, and some writers have considered that the deposits accumulated in even the delta of the Mississippi would have taken at least 100,000 years. The uncertainty of the system of estimating how long it ought to have required to accumulate a deposit during former times, by noting what now occurs, must leave us in great doubt as regards our results, because we do not, from geological investigation, know all the forces formerly at work.

In Sir C. Lyell's *Antiquity of Man*, page 43, he states: 'I have shown in my travels in North America, that the deposits forming the delta and alluvial plain of the Mississippi consist of sedimentary matter, extending over an area of 30,000 square miles, and known in some parts to be several hundred feet deep.

Although we cannot estimate correctly how many years it may have required for the river to bring down from the upper country so large a quantity of earthy matter—the data for such a computation being as yet incomplete—we may still approximate to a minimum of the time which such an operation must have taken, by ascertaining experimentally the annual discharge of water by the Mississippi and the mean annual amount of solid matter contained in its waters. The lowest estimate of the time required would lead us to assign a high antiquity, amounting to many tens of thousands of years (probably more than 100,000), to the existing delta.'

The method of estimating time by the thickness of various deposits, is that which has been adopted by this eminent geologist, and also by many other writers. If we were certain that the conditions under which the deposits were accumulated were formerly exactly the same as they are now, this method would be quite sound, and the conclusions based thereon probably very near the truth. When, however, the climatic conditions shown by geology to have existed formerly are very different from those now prevailing, we at once have presented to us an uncertainty relative to the forces at work, and we can have no clue as to the annual amount of the deposits made under climatic conditions entirely different from those which now act.

When we bring astronomy to our aid, and find that the present course of the pole, if continued and traced back to the past, gives us 17,000 winters, during a part of which time the arctic circle came down to about 54° latitude, and thus allowed of vast accumulations of snow and ice in the delta of the Mississippi, exceeding many times the greatest amount ever now seen there, and that these winters alternated with 17,000 summers, during a portion of which the sun attained a midday altitude only 18° from the zenith of 54° latitude, and remained at midsummer twenty-four hours above the horizon, thus melting and scattering not only the snow and ice accumulated in this delta, but even the ice and snow round the pole, we find that one such winter and summer must have produced more deposits than fifty winters and summers under present climatic conditions.

Thus, to estimate geological time by the thickness of deposits made under former conditions, when we do not know what those conditions were, must leave us in considerable uncertainty. We believe, therefore, that when those distinguished geologists who have made these estimations of the time required to form deposits, realise the fact, that the course traced by the earth's axis on the sphere of the heavens requires no theory or speculation to account for changes, but merely requires to be traced back exactly as it is now moving, when it will give us a climate which must have annually produced far greater deposits than now occur, they will immediately modify their opinions relative to the time required to produce the observed facts.

Mr. Prestwich, in his able article already quoted (see *Philosophical Transactions*, 1864), refers to the same fact. Speaking of time, he says: 'The next possible standard of measure is the time required for the excavation of the valleys themselves. I have already described the agents which probably co-operated in this gigantic operation. That it must have been one of great time there can be no doubt; but like the operations at present in progress, by no means furnish us with the gauge of the rate of the denuding action. In considering this point, there is, besides the greater floods and severer cold, another element which must not be overlooked. This is the varying solvent power of spring and river water. . . . I do not, however, feel that we are yet in a position to measure that time, or even to make an approximate estimate respecting it. That we must greatly extend our present chronology with respect to the first existence of man appears inevitable; but that we should count by hundreds of thousands of years is, I am convinced, in the present state of the inquiry, unsafe and premature.'

Even the estimation of time formed from an investigation of the growth of coral reefs must be uncertain, for we know not how rapidly these might progress under the climatic changes formerly occurring.

A geological question requiring far more delicate handling than the rate of deposits, and likely to give us far more accurate results than can be obtained from their examination, is, the investigation whether more than *one* glacial epoch occurred, producing *similar* effects at different times.

From an examination of the course traced by the pole of the heavens, it appears that at about 13,000 B.C. the conditions of the glacial epoch were at their height. If the course of the pole were uniform, and either undisturbed by any other cause, or only slightly disturbed, another glacial epoch would have occurred and have been at its height 31,600 years previously, that is at about 44,000 B.C.

This problem is one mainly depending on geological evidence, although the astronomical evidence tends to demonstrate that several such epochs may have occurred, and therefore that between the present date, and the pliocene, probably two or three revolutions of the pole, each of 31,000 years, occurred.

The evidence would mainly depend on the relative elevations of land, as shown by the glacial deposits.

Professor Ramsey, from his investigations in North Wales, concludes that there must have been three successive glacial epochs, during which considerable change occurred in the elevation and depression of land. If three such periods did occur, the height of the last would have been about 13,000 B.C., that of the second 44,000 B.C., that of the earliest 75,000 B.C.

Another method which has been adopted by geologists to arrive at some date for geology, is to note the amount of upheaval which is now taking place in various parts of the world, and then by proportion to estimate how many years it would require to raise land as much as some parts of England or Wales are known to have been raised since modern geological times.

This method has the same defect as that brought forward relative to deposits, viz. the conditions now are not what they were, and the comparison is therefore unsound.

At a time when the earth was subjected to an arctic circle down to $54\frac{1}{2}°$ and tropics up to $35\frac{1}{2}°$, the changes annually from heat to cold must have been immense: are we warranted in assuming that these changes of climate had nothing to do with the elevation or depression even of mountains?

The terrific rush of waters each summer, caused by the rapid melting of ice and snow, would in itself produce denudations and deposits of immense amount, especially when continued 17,000 years; and it seems probable that it is during these periods of great alternation of climate that the greatest upheavals and depressions occur, and not during such uniform or quiet periods as those which have prevailed during the past 8000 years.

When an examination of the changes which are now occurring on earth proves that during some 2000 years these changes are but slight, we naturally begin to reason from this long period and its comparative uniformity, and hence to assign enormous epochs of time for the date when the mammoth roamed over

the land, and endless ages for him to become extinct.
But, again, to form this estimate we omit to give due
weight to those powerful agencies which were at work
during the glacial epoch. If, as is most probable, the
mammoth became extinct during the severe cold of
the winters of the glacial epoch, it would not have
taken so very many of these winters to exterminate
him. It must also be borne in mind that at that remote date, man would have been a resident in tropical
regions only, and the mammoth consequently would
have been hunted down and killed had he entered
the thickly-peopled districts of the tropics. In many
parts of Europe also his migration south would be
limited by the seas; such most probably was the case
in England; thus he had the choice of encountering
one of two enemies—man or frost—either of which
would cause these giants to rapidly disappear. Hence
we may bring the mammoth period nearer our own
time, and man nearer our own time, than those
hundreds of thousands of years which have been assumed by some writers as absolutely necessary. Mr.
Prestwich has also, from geological investigations,
come to a similar conclusion, as shown in the following extract:

'Whilst abstaining from any general hypothesis
in explanation of the phenomena, there is, however,
one point to which I must refer before concluding,
although I cannot at present venture beyond a few
generalities respecting it. It might be supposed that
in assigning to man an appearance at such a period,

it would of necessity imply his existence during long ages beyond all exact calculations, for we have been apt to place even the latest of our geological changes at a remote and, to us, unknown distance. The reasons on which such a view has been held, have been mainly the great lapse of time considered requisite for the dying out of so many species of great mammals; the circumstance that many of the smaller valleys have been excavated since they lived; the presumed non-existence of man himself; and the great extent of the later and more modern accumulations. But we have, in this part of Europe, no succession of strata to record, a gradual dying out of the species, but much, on the contrary, which points to an abrupt end, and evidence only of relative, not of actual time; while the recent valley-deposit, although often indicating considerable age, shows rates of growth which, though variable, appear on the whole to have been comparatively rapid. The evidence, in fact, as it at present stands, does not seem to me to necessitate the carrying of man back in past time so much as bringing forward of extinct animals towards our own time; my own previous opinion, founded on an independent study of the superficial drift, or pleistocene deposits, having likewise been certainly in favour of the latter view. There are numerous phenomena which I can only consider as evidence of a sudden change, and of a rapid and transitory action and modification of the surface at a comparatively recent geological period —a period which, if the foregoing facts are truly interpreted, would seem nevertheless to have been marked before its end by the presence of man, on a land clothed with a vegetation apparently very similar to that now flourishing in like latitudes, and whose waters were inhabited by testacea, also of forms now living; while on the surface of that land there lived mammalia, of which some species are yet the associates of man, although accompanied by others, many of them of gigantic size, and of forms now extinct.'*

Whilst, however, it thus appears that we somewhat curtail the allowance of time between the present and the glacial epoch, yet there seems to be a vast number of revolutions of the equinoxes to produce the endless seams of coal with their intermediate beds of sandstone and shale. Thus, for example, between the Sharp and Broad mountains there are sixty-five seams, each of which is separated by what we may term a glacial epoch, or period of 31,000 years, thus making a period of above 2,000,000 years for these alone. When we consider what there is between the coal-measures and the boulder period, there is time in plenty for those who insist on these vast epochs.

Sir C. Lyell, in *Antiquity of Man*, page 285, calculates that the submergence of Wales to the amount of 1400 feet would require 56,000 years; but if 800 feet more submergence took place, an additional 32,000 years would be required, amounting in all to 88,000 years.

* *Philosophical Transactions*, 1860.

But this result is arrived at only by assuming 2½ feet per century to be the rate at which this submergence occurred. If the rate were greater, as it probably would have been when the climatic conditions were very different, the number of years would be very different. As this distinguished writer admits, the rate is purely a conjectural one; and as on the coast of Spitzbergen the rise at present is at least six feet a century, one-third only of the 88,000 years would be required to account for the rise of the Welsh hills, at the rate at which Spitzbergen is now rising.

The question relative to the antiquity of man is one intimately connected with the date of the glacial epoch, consequently it is a matter of deep interest whether it was 14,000 or 88,000 years ago that the boulder period occurred.

When it is evident that, from geological investigations, we can arrive at an approximation to the time at which the glacial epoch occurred only by conjectures relative to the rate at which land now rises or falls, or deposits are now made, we see how uncertain must be our conclusions.

A movement, however, of the whole earth, which movement has continued, as far as history carries us, not only uniform in rate, but uniform in direction, appears to give us far more certain information as regards a former date than we can obtain from deposits, or from a rise and fall of land.

When we find that during 2000 years the course

which the pole has traced relative to the pole of the ecliptic is the arc of a circle, which circle has for its centre a point 6° from the pole of the ecliptic, we naturally inquire what causes are there that should produce any other movement than that which any competent geometrician can recognise as the course which has been traced — and we can find none. During the past 2000 years the various planets have been in every possible position relative to the earth, and a complete cycle of changes has been effected, and yet astronomical records supply ample evidence that the movement of the pole has been uniform. Thus we have sufficient grounds for giving the date 13,000 B.C. as about the time at which the last great boulder period was at its height.

If we were to even refer to all the views which have been brought forward during the last few years relative to the time required for the various deposits of sand, clay, gravel, &c., we should occupy a larger space than is given in this work. We must therefore be content to limit our remarks to a few general principles.

When attempts have been made to estimate the length of time occupied in forming various deposits, and the data on which this calculation has been attempted are the thicknesses of deposits which now take place during a century, a very close approximation may be arrived at, if we do not attempt to estimate more than 10,000 years from the present time. Going back to about 8000 B.C., we find the pole then at an angular distance from the pole of the ecliptic of about 29°; a condition which, although likely to increase the annual deposits by the alternation of the summer and winter temperatures, would yet cause but trifling differences. When, however, we enter the boulder and drift period, we believe every reader will at once perceive that when the obliquity was 35° 25', the deposits during one century would be far greater than during a hundred centuries under present conditions, and therefore our proportion, or rate, fails to give us even approximate results if we go beyond the last indications of the drift. Thus the conclusions of M. Morlot relative to the dates of the bronze and stone age may be very nearly correct.[*] M. Morlot estimates, from the amount of deposits found at various places where evidences of man's antiquity exist, that the bronze age occupied about 2900 years, and the stone period from 4700 to 7000; for the two periods about 7400 or 11,000 years were required. Thus at the date about 6000 to 9000 B.C., Switzerland had its lake dwellers, and Abbeville its users of stone implements; and the climatic conditions were then so very slightly different from those now prevailing, that Europe would have been quite fit for a race of hardy hunters.

That which we wish to particularly point out, however, is, that to estimate dates by the thickness of deposits made *during* the glacial epoch, on the assumption that they then occupied, foot for foot, as long a

[*] See *Prehistoric Times*, p. 319.

time to accumulate as they now occupy, will lend to serious errors. As well might we estimate that because the obliquity now decreases 45″ per century, therefore it would take 96,000 years to decrease 12°, a result which would be erroneous. *Causes* in both cases must be looked to.

When it appears probable that man existed on earth during the glacial epoch, and it also appears that the height of this epoch occurred about 15,000 years ago, it becomes interesting to inquire what chance there is of some records of this event, as connected with the altitude of the sun, having been handed down to us from ancient writers.

On examining the course traced by the earth's axis in the heavens, we find that at a date about 6000 B.C., the sun's declination varied annually from about 29° south to 29° north. This variation would scarcely have produced any changes in the climate of Egypt, Central India, or Asia, different from that which now occurs; in fact the climate of all localities within about 30° of the equator would be almost similar to what it now is, even with an obliquity of 35°,. the great changes occurring only in high and middle latitudes.

Any human records from *actual observation*, with regard to this change of the sun's declination, ought then to come from a date about 8000 years ago, and from people who lived in or visited high northern regions. It is unnecessary to remind the reader that no such records exist; consequently the only possible manner in which any history of the former conditions could reach the writers of even 2000 years ago, was in the form of a tradition handed down from a date at least 4000 years previous to their time. How very improbable it is that any true or accurate account of the conditions which really prevailed would be passed from generation to generation during 4000 years; but that some vague reports should have been received, and some vague superstitious belief still maintained among the eastern nations, seems probable. It must be borne in mind also that 2000 years ago the rate in the decrease of the obliquity would have been much greater than it is now. During the present century the rate is only 45''; but at the date 1000 A.D. it would have been 130'', if the pole moved relatively to the pole of the ecliptic as it now moves. So also in ancient times it would have been much more rapid, and therefore more likely to attract the attention of the ancients. Thus any vague belief or report of the ancients relative to changes of climate, or to the sun's position, becomes of interest.

In Wilson's *Chronology of the Hindus* it is mentioned that these people have a tradition that the world is renewed at the commencement of each menu or period of 12,000 years, that at these periods there is a deluge, and that the present is the seventh. The author of the Persian manuscript Modjmel also ascribes to Zoroaster a similar statement.

Proclus in his treatise *De Sphæra*, cap. i., states

that the Chaldeans had an opinion that there was a great planetary period during which the universe experiences an entire revolution. They stated that the commencement of the world coincided with the beginning of this period.

Aristotle also refers to this belief among the Chaldeans, and calls the period 'a great year.' He gives the following additional and interesting information: During the winter of this period there is a *cataclysmos* or deluge, and at its summer an *ecpyrosis* or conflagration.

A revolution of the pole of the heavens would produce the same effect as an entire revolution of the universe, and when the two poles were $35\frac{1}{4}°$ apart, and the sun was at midday vertical at latitudes of $54\frac{1}{2}°$, and did not set at midnight, and thus, like one of the stars of the Great Bear, continued above the horizon during the whole of its course, and this for days together, something very like an *ecpyrosis* would occur. When also the ice and snow, accumulated during the winter of this great period, were suddenly thawed by the spring sun, a condition would be produced not badly described as a cataclysmos.

The statement of the Egyptians to Herodotus relative to the sun having changed its course during 11,340 years may have some foundation in truth, and if it referred to the change in the sun's declination of $12°$, it would be quite true, as they stated, that it produced 'no alteration in the climate of Egypt.' For

Egypt is so near the tropics, that a variation of even 12° in the sun's declination would produce but little climatic change in that country.

Plato also relates a somewhat similar incident to the preceding, showing that the Greeks had some such tradition among them.

To expect that with the imperfect state of literature and science possessed by the men of the East some 3000 B.C. we should obtain any correct records of the sun's declination, would be expecting too much. If man were a denizen of earth when the climatic changes in middle latitudes were annually as great as they would be when the two poles were $35\frac{1}{4}°$ apart, it is probable that some vague tradition of this climate and change should have been handed down, even during the long period of some 14,000 years. Thus, although it would be somewhat rash to base a conclusion on these traditions, yet when we find that they existed among the Hindus, the Persians, the Egyptians, and the Greeks, we should hesitate before we reject all such statements as superstitions, or as a proof of ignorance.

As before remarked, the climate of the earth north of 40° latitude would not have been suitable for men during the glacial epoch. The changes between summer and winter would have been as great as between an arctic winter and an excessively hot tropical summer. Man would naturally emigrate from a country where he was frozen up in winter, and almost burnt up in summer. Thus, if man has been, as is supposed by some geologists, a resident on earth for many tens of thousands of years, we are more likely to find his remains in middle latitudes, such as England and France prior to the glacial epoch, viz. in the sand and clay of the pliocene, when the terrestrial climate was more like what it now is, than amidst the drift or erratic phenomena.

If, as is supposed by M. Morlot, the Swiss lake-dwellings date back only about 7000 years, these would have been in existence just when the glacial epoch had terminated, and when the European climate was more uniform than during the height of the cold and hot period. Probably these lake-dwellers were the first human beings who migrated from the tropics, and located themselves on the borders of those waters where the remains of their dwellings are now found.

CHAPTER X.

OUTLINES OF SOME OF THE RESULTS OF THE CURVE TRACED BY THE EARTH'S AXIS.

Having found from the experience of some years that either no really formidable objections could be brought, or were brought, against the movement of the pole being such as we have described, we will now state what are the advantages derived from a knowledge of this movement.

First, we are enabled to trace the course of the pole in the heavens by aid of a geometrical problem, instead of admitting into astronomy either a geometrical impossibility, or allowing the course of the pole to be vague, and undefined, when dealing with long periods.

Secondly, we can give the true definition of what the course of the pole really is, and thus obtain the value of the obliquity by a geometrical formula, each step of which is intelligible, instead of by observation only, or by an empirical rule which fails except for very short periods.

Thirdly, the same law which enables us to obtain astronomical data by calculation, hitherto only obtained by observation, gives a full and ample cause, and gives us the exact date of a great geological event or period in the earth's history. The facts of geology tend to verify the true movement of the earth's axis at a date far beyond that at which we possess human records; and whilst we find that during hundreds of years recorded observations tend to confirm the course of the pole as the one we have described, geology gives evidence as regards ten thousand years ago.

Fourthly, there result from this movement of the earth's axis other effects of a most important nature, which explain those few and trifling discrepancies known only to those who have long studied astronomy. Thus the calculation of certain events with great accuracy can be as readily accomplished, when we know the true polar movement, as can the value of the obliquity for any date, although hitherto these events have been correctly obtained only by long and laborious observation, and even then require continued observation, in order to assign correctly the proper values for the future.

These appear to be results of sufficient importance to justify calling attention to the problem here briefly referred to.

The following formulæ will aid to explain how various interesting data may be obtained:

First, for the obliquity at any date, past or to come:
$$[2298 - T] 20''\cdot 05 = a.$$
$$\frac{a}{\sin. 29°\,25'\,47''} = C \qquad \cos. C \tan. 6° = \tan. B$$
$$\cos. O = \cos. 6° \left\{ \frac{\cos. [29°\,25'\,47'' - B]}{\cos. B} \right\}$$

Where T is the date before or after 2298 A.D., O = obliquity.*

---

* The reader is recommended to test for himself the truth of this formula, by comparing the results obtained by it for any year with the

VALUE OF THE OBLIQUITY BY CALCULATION.

From this formula the following are the values obtained for the obliquity, and the rate per century for 1300 years:

YEAR A.D.	OBLIQUITY.	RATE PER 100 YEARS.
2300	23° 23′ 47″*	0″
2200	23 25 52	5
2100	23 26 7	15
2000	23 26 33″·5	26
1900	23 27 10	36
1800	23 27 56	46
1700	23 28 53	57
1600	23 30 0	67
1500	23 31 17	77
1400	23 32 45	88
1300	23 34 23	98
1200	23 36 11	108
1100	23 38 9	118
1000	23 40 19	130

At the date 6000 B.C. the obliquity would have been about 29° 25′, and at this date the glacial epoch was just terminating. At the date 13,000 B.C. the obliquity would have been 35° 25′, and the glacial epoch would have been at its height.

The value of the precession for a polar movement of 20″·05 may be obtained approximately or for a short period by dividing 20″·05 by the sine of the obliquity. But to obtain this value accurately or for long periods, we must find this value in another way, viz.

Let s be the pole at any date, say 1400 A.D., R the

---

recorded value of the mean obliquity of the ecliptic for that year in the *Nautical Almanac*.

* In consequence of our log-tables giving only seven places of decimals, our results can only be given to seconds with certainty.

pole at any other date, say 1800 A.D., the arc R S is traced at right angles to S C, not at right angles to S E; consequently the angle R E S will represent the precession between the dates 1400 and 1800 A.D. In consequence of the polar movement from S to R being traced at right angles to the arc joining S and C, therefore E S R is less than a right angle, and the polar movement is directed towards a meridian short of that of 0 hours R.A.

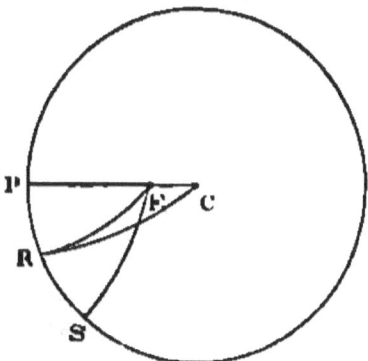

At the date 2298, when the pole is at P, the polar movement will then be directly at right angles to the arc joining the pole P with E, the pole of the ecliptic.

The angle E C R, E C S, &c. we term the eccentricity of polar motion. This angle will be greatest when C E is at right angles to the arc joining C and the pole of the heavens. The value of this angle at its greatest value will be $\dfrac{6°}{\sin. 29° 25'} = 12° 5'$. The value of this angle at any other date, such as when the pole is at s, may be approximated to by the equation, 12° 5′ × sin. of angle P C S = E C S; consequently when the pole is at P this angle is at zero.

Most important results connected with the measurement of time, also connected with changes of stellar right ascensions, follow the fact of the point c being the centre of polar motion, and not E the pole of the ecliptic.

In order to clearly understand the effects of a complete revolution of the pole of the heavens round the pole of the ecliptic, the reader unacquainted with the higher branches of geometrical astronomy should study with care chapter XVIII. of Sir J. Herschel's *Outlines of Astronomy*, particularly article 909. He will then understand how, during a whole revolution of the equinoxes, the meridian will have lost one exact rotation on a star; so that a star *within* the circle described by the pole passes a meridian once oftener during a revolution of the equinoxes than does a star outside this circle. When, then, this circle is supposed to have a radius of 23° 28′, and to have for its centre the pole of the ecliptic, whereas it appears from the course traced by the pole that the radius is 29° 25′ 47″, and the centre situated at a point 6° from the pole of the ecliptic, it follows that most singular confusions and perplexities must arise as regards the changes in stars' right ascensions.

From a combination of the earth's diurnal rotation and the revolution of the pole of the heavens in a circle around *some* point as a centre, it follows that the only uniform measure of time during one entire revolution of the equinoxes must be the interval between the successive transits of this centre of polar motion, and as the pole of the ecliptic is not the centre of polar motion, it follows that the successive transits of that pole are not uniform measures of time. An exact uniform measure of time will be obtained by taking the interval between the successive transits of that point in the heavens which is the centre of polar motion, and which appears to have a north polar distance (constant) of $29° 25' 47''$, and a right ascension at the present time (1872) of about 17 h. 45 m.

Hence, when we compare the right ascensions of stars given in a catalogue at some former period with their right ascensions given in a later catalogue, and find how nearly every star exhibits such a change of rate during this interval as is assumed to be a 'proper motion,' we cannot avoid noting how futile will be any continuous labours in this direction, based on calculations which assume that the course traced by the earth's axis is a circle, from the centre of which circle it is secularly varying its distance.

When we consider what results follow the tracing out of this curve, we believe their importance will be appreciated by every competent geometrician and practical astronomer. To obtain minutely-accurate observations relative to a limb of the sun is excessively difficult, because the atmosphere on a bright sunny day is usually very tremulous, and the heat of the atmosphere varies so much at different heights, that the allowance to be made for refraction is always somewhat uncertain; and thus, to determine with minute accuracy the obliquity is attended with difficulty. Of course any astronomer must know that the rate at which the obliquity of the ecliptic should vary would be *almost* uniform for a few years; and therefore he can check his observations, and avoid any palpable error; still there is some uncertainty, in consequence of the variableness of refraction. Thus we find that $0''\cdot 457$ is assumed to be the uniform annual rate of the decrease in the present century; an assumption which is not geometrically true; for if $0''\cdot 457$ were the rate at the commencement of the present century, it could not be the rate at the end, because a curve cannot approach uniformly a point, unless the point is in the direction of the curve.

When we examine the recorded mean obliquity in the *Nautical Almanac*, we find that the rate $0''\cdot 457$ is not always adhered to: thus the mean obliquity for 1861 is given for the 1st January $23° 27' 26''\cdot 92$, and for 1865 the mean obliquity on the 1st January $23° 27' 24''\cdot 69$; the difference is $2''\cdot 23$, which, divided by 4, the interval in years between these dates, gives $0''\cdot 557$ per annum, or $55''\cdot 7$ per century, if the rate were uniform.

Thus, to be enabled to obtain this important item in astronomy with minute accuracy, and independent of observation, is of course most desirable; and this result seems to have been achieved by the geometrical formula we have given.

If we treat this subject with minute accuracy, we must state that the decrease in the obliquity cannot be exactly the same for any two years in succession; so that when a constant quantity is subtracted each year, the error accumulates, and at length a fresh adjustment, like that mentioned above, has suddenly to be made, and another erroneous addition continued until another adjustment is made, and so on.

In order to obtain the value of the obliquity with accuracy for any date, we may work out by our formula the exact obliquity for each tenth year, and then interpolate for the intervening years. Adopting this method of calculation for centuries, it will be found that between 1800 and 1900 there will be a decrease of $46''\cdot5$, which will give a mean rate for the whole century of $0''\cdot465$ per annum; but this mean rate will not be the annual rate either at 1800 or 1900 exactly, but will be the rate at about 1850. As we advance towards the year 1900 the rate will be found less than $0''\cdot465$ annually, and consequently less than $4''\cdot65$ in ten years. It will be interesting to note when this fact will be discovered by observers; but it is evident that as the decrease in the rate will be but slight, we must have more minute accuracy in our observations in order to discover it, than is shown by the above-quoted recorded change from $0''\cdot457$ to $0''\cdot557$ per annum.

The following calculations may aid those who

wish to note the decrease in the rate per century at
which the obliquity has and will decrease:

From A.D. 2000 to A.D. 1900 . . . 36"
„   1900 „ 1800 . . . 46
„   1800 „ 1700 . . . 57

Thus we should have to interpolate for second differences, in order to obtain minute accuracy, if we required the exact rate for any one year.

Let us now refer to what we are enabled to accomplish when we have correctly localised the centre of polar motion, and have defined the curve which the pole pursues with that regularity which marks all celestial movements.

We can predict the exact position of the pole for any time, either in the past or in the future. The calculations which we have given demonstrate that, starting from a period above 400 years in advance, we calculate back to the present time, and arrive by this calculation at results accurate to within half a second. But the pole of the heavens is, after all, only a point in the heavens; and we can localise this point with minute accuracy for all time. Need we point out to the geometrician *that every star in the heavens is but a point*, and that the position of these stars can likewise be localised by a calculation as simple as is that which enables us to find the obliquity? for we can substitute a star for the true centre of polar motion, and find the distance of the pole from any star, and hence the star's polar distance for any time, just in the same manner as we find the other item. So also can the star's right ascension be calculated for any short or long period in advance; the key to the solution of this problem being a knowledge of the true centre of polar motion —a knowledge not hitherto possessed by any astronomer.

We fear, however, that such a result will not be very popular among some toilers in the astronomical labour-market. Just as the skilled mechanic detests the machine which does well in one day as much work as he accomplishes imperfectly in a month, so it is most probable that those careful observers and computers who now by scores are employed in registering transits of stars, and in working out minor computations, will not lavish their affection on a geometrical calculation which arrives in ten minutes at a result which they can only attain in as many months. When also we find how elaborate have been the tables formed of stars, all of which have been supposed to have a very large amount of 'proper motion,' it seems, indeed, unfortunate that for these words 'proper motion' we must in many cases substitute 'results of erroneous principle of determining stars' positions.'*
It is a fact, however, that we might with equal justice attribute to the pole of the heavens a 'proper motion' because it varies its distance from the pole

_{* We do not mean to state that *no star* alters its relative position at all; but we do find that very many stars which have assigned to them a large 'proper motion' have this alteration of position almost entirely in consequence of the pole and the meridian moving in a manner different from what has been supposed.}

of the ecliptic, as to attribute to every star a proper motion because it varies its distance from *our zenith* and the pole in a manner that is not in accordance with our theories.

We use the term 'zenith' advisedly, because there are items depending on the movement of our zenith, resulting from this polar motion, of an importance which cannot be overrated.

Thus the practical astronomer will probably believe that the geometrical facts here brought forward are not likely to benefit his position. We can, however, assure him that there is far more yet to be done in practical work than has ever hitherto been done; the difference being the direction in which his labours should be turned.

CHAPTER XI.

CONCLUDING REMARKS.

We have in the preceding pages endeavoured to lay before the reader in as brief and clear a manner as possible a problem of great importance. This problem is one to which many inquirers have devoted considerable attention; but we believe no one has ever before trodden the same ground that we have. That the real course traced by the earth's axis was at the present time anything but a circle, the centre of which circle was either the pole of the ecliptic or close to it, seems not to have been suggested as even possible. Yet when it was known that there was a secular decrease in the obliquity, this fact might have been noticed; and the next step should naturally have been to inquire where was the centre of this circle.

The reason why so little attention appears to have been devoted to this special problem by modern inquirers is manifest. First, geometry at the present time is not as popular as is physical astronomy. The laws relative to the attraction or repulsion of bodies, those referring to masses of planets, the probability of some remote star having in its constitution hydrogen or thallium, or some other metal, the variation in the sun's atmosphere, &c., are subjects to which hundreds of individuals devote their time and thought; whilst the dry, hard, yet rigid facts connected with a geometrical investigation relative to the course traced by the earth's axis is so unattractive, that this problem has been allowed to rest for many years.

It seems highly probable that very many persons have been prevented from inquiring on this matter in consequence of the belief that the problem was exhausted. The great reputation of M. Laplace would naturally produce this effect upon those unacquainted with geometry. The fact that M. Laplace announced that the plane of the ecliptic could only vary to the amount of 1° 21′, and hence that the variation of the obliquity was confined within very narrow limits, would probably be the conclusion of those who omitted to notice that a variation in the plane of the ecliptic could produce no variation at all in the obliquity, provided it were true, as stated, that the pole of the ecliptic was the centre of polar motion.

So completely impressed do many astronomers appear to be with the idea that the only possible cause of a variation in the obliquity must be a variation in the position of the *plane* of the ecliptic, that not long since a popular astronomer stated as a fact, that as M. Leverrier had decided that the plane of the ecliptic could only vary to the amount of 4° 52′,*

---

* There appears to be some singular confusion among various writers relative to this variation in the plane of the ecliptic. Those who copy

'hence it follows that the greatest possible distance of the tropics from the equator is 28° 20'."

It is, we trust, unnecessary to again point out, that unless the exact centre of polar motion be known, such statements as the above are scientific absurdities; for a variation of 12° in the obliquity must occur even with no variation in the plane of the ecliptic, if the earth's axis continue to trace the same course in the heavens which it has traced during the past 1800 years.

When it was found that the facts of geology were of such a nature as to require an explanation why so singular a climate as that evidenced during the glacial epoch occurred formerly, but does not now exist, speculations of various kinds were advanced as regards the probabilities of catastrophes on earth, such as vast waves, earthquakes, &c. Such speculations

---

from M. Laplace still state that 'the plane of the ecliptic cannot possibly vary more than 1° 21'.' Others state that the plane cannot possibly vary more than 3° (see Arago's *Popular Astronomy*, vol. ii. p. 899 and note, relative to Laplace's investigations). Others again, who refer to M. Leverrier's more recent mathematical theories, give as much as 4° 52'. So that it really appears as though this variation were gradually becoming more elastic. As, however, the plane of the orbit of the planet Venus varies only 3° 23' from the plane of the ecliptic; and as in Venus the arctic circle comes to within 15° of her equator in winter, whilst a vertical sun, and hence the tropics, reach to within 15° of her poles in summer,—it will be manifest that the variation in the plane of the ecliptic does not limit the extent of the arctic circle, but is dependent on the position of the axis of the planet. Those who would like to work out a grand glacial epoch should study the annual climatic changes occurring in Venus.

were very wisely opposed by the most careful and thoughtful philosophers of the day, who urged, that nature always appeared to act uniformly and gradually, and rarely by convulsions. To such reasoners we believe this polar movement will present no difficulties.

Instead of searching for unusual catastrophes to explain known effects, we have merely to examine the recorded astronomical evidence of the past 300 years. We find that the earth's axis has traced a curve in the heavens during all time of which we have records. We find that the point which traces this curve was in 1871 at a distance of 23° 27′ 22″ from the pole of the ecliptic, that it moves laterally as regards the pole of the ecliptic about 20″·05 annually. At 1840 the curve was 23° 27′ 36″·5 from the pole of the ecliptic; at 1795 it was 23° 27′ 57″·6; at 1690 about 23° 27′ 4″·8.

What, then, is this curve, and from whence did the pole come that is tracing it? We have merely to continue this curve just as it is now tracing its course in the heavens, to find that it came from a point in the heavens distant 35° 25′ 47″ from the pole of the ecliptic, at a date about 13,000 years B.C. No speculations or probabilities are necessary or required; we have to treat an abstract fact only, viz. what is this course traced by the earth's axis, and we find it a circle, the radius of which is 29° 25′ 47″.

Just as two or three observations on the position of a comet enable us to calculate the course of that comet, so do these polar positions enable us to calculate the positions and course of the pole.

By merely taking as dependable the recorded astronomical observations relative to the position of the pole, we find that we can calculate by a formula (based upon the circular movement of the pole around a point 29° 25' 47" from it) the exact position of the pole relative to the pole of the ecliptic for all dates; also we find that this same course of the pole must have produced on earth about 13,000 B.C. such a singular climate, that the evidence of this climate ought to be even now manifest.

Thus, if geology had not been a science, and if no observers had collected the evidence connected with the glacial epoch, yet the course traced by the earth's axis must, when correctly understood, have led geometricians to search for such corroboration of this course as ought to be found on the earth's surface. Fortunately the labours of geologists have already supplied this evidence, and have to a great extent proved that some such condition did formerly occur; when, or by what means, has hitherto remained a mystery. A few distinguished men, in answer to our request that they would criticise and find every possible fault with this problem, replied, 'That they would rather hear what X or Y said.' 'That they did not like to offer an opinion on the subject.' 'That they had asked the advice of a very competent astronomer, who assured them everything was known about the pole.' 'That they were so occupied about something else (very important), that they really had not the time to examine the matter.' 'That we ought at once to submit the problem to Professor A or B, who would give it every attention.'

'On submitting the problem to A or B, we usually received the reply, that 'if we looked in any standard work on astronomy, we should find that the pole of the heavens traced a circle round the pole of the ecliptic as a centre, and never varied its distance of 23° 28' from this centre; and *therefore* the pole did not move, as we thought it did.'

During the investigation of the problem here brought forward we have been often much surprised to find how many individuals who, one would expect, ought to be interested in this matter, avoided all examination, and even refused to look at this inquiry, on the plea that they 'were occupied about some important astronomical matter, and could not spare the time.'

To make photographs, sketches, and notes of the red flames during a total eclipse, to examine what lines become visible in the spectrum during the same phenomenon, and to speculate on the cause of the sun's spots, are matters of great interest and importance, and we are justified in spending thousands on an expedition for the purpose of making researches on these subjects. So also to decide whether the sun is 91 or 95 million miles from the earth, is an interesting problem justifying years of preparation for a transit of Venus, and hundreds of thousands being

spent in various countries for the purpose of fitting
out expeditions to investigate this problem. Can
these two astronomical events, however, be compared
in importance and in their practical use, with the
problem of deciding on the true movement of the
earth itself? Surely it cannot be claimed that it is
more important to know whether sodium, or hydrogen, exists in the sun's atmosphere, than it is to discover a movement of the earth, which reveals the
cause of our last geological change, enables us to calculate the obliquity of the ecliptic for any number of
years, and also leads to our being able to calculate
the position of every star with equal ease.

How many thousand of observers have been employed during the past hundred years in finding the
obliquity alone, and how many are still so employed!
Yet this item can be calculated to within a tenth of
a second. How many observers and computers are
now occupied in endeavouring to correct those star-
positions which cannot now be calculated, because
the pole of the heavens traces a course which has not
hitherto been correctly understood!

When, then, we find that individuals plead that
they have important problems in astronomy occupy-
ing them, and therefore cannot spare time for inquiry, it reminds us of the conditions of 2000 years
ago, when the ancients were discussing the relative
influence of Mars or Saturn on the career and fate of
nations, and had not the time to devote to anything
so unimportant as to inquire whether the earth was

round or a flat surface, or whether it rotated on its axis, or was stationary.

These experiences, and many others of a similar nature, have led us to conclude that there is considerable difficulty in inducing certain persons to bring to bear on any novelty, those identical mental powers which have alone rendered their names well known to science.

It takes a long time to render us familiar with a new idea, and the history of the past shows us how slowly any new truth ever made its way.

Fortunately, however, the problem which we have here dealt with is not one depending on opinion or probability; it is not a question as to what is likely, but it is a question of so purely a geometrical nature as to be capable of demonstration.

We have a curve to deal with, and we have to define what this curve is. It has been hitherto stated to be a circle, the centre of which is the pole of the ecliptic. We suggest that this statement contains a geometrical impossibility when we admit a decrease in the obliquity.

It is then stated to be 'a circle for a short time, but not a circle for a long time;' a definition as contradictory as is the former.

We suggest that the curve is part of a circle having for its radius an arc of 29° 25′ 47″, the centre of this circle being 6° from the pole of the ecliptic. This curve is demonstrated by observation to be the true curve. It explains geological facts hitherto unexplained, and enables us to calculate astronomical data hitherto obtained merely by observation; and as such we submit it to the investigation of those who may be interested in the matter.

It may not be uninteresting, after the investigations which have been made, and the objections that have been answered, to explain how the present accepted movement of the pole of the heavens in a circle around the pole of the ecliptic found its way into astronomy.

In the system promulgated by Copernicus[*] he supposed the earth to have three movements. First, its diurnal rotation; secondly, its annual revolution about the sun; thirdly, a conical revolution of the earth's axis about an imaginary line perpendicular to the plane of the ecliptic, and passing through the earth's centre, which he imagined to be the vertex of the cone. The inclination of the earth's axis to this line was supposed equal to the mean obliquity of the ecliptic. In order to account for the retrograde movement of the equinoctial points, he supposed that the conical revolution of the axis was completed in 20^m less time than a revolution round the sun; this 20^m converted into an arc representing $50''$, the value of the precession.

This conical movement of the earth's axis was shortly after the time of Copernicus accepted as the cause of the precession; but it was then agreed that it was not an annual motion, but one of long dura-

[*] *De Revolutione*, lib. i. cap. 2.

tion. As, however, it was not then known that the obliquity of the ecliptic secularly decreased, or because no geometrician then pointed out the discordance, it was also decided that the earth's axis still traced an exact cone round an imaginary line perpendicular to the plane of the ecliptic.

When, then, it was supposed that some few stars indicated a change of latitude, and that observations proved that there was a decrease in the obliquity, it seemed a ready and simple explanation of both to assign a change to the plane of the ecliptic. Thus, when a great many theories have been for years taught as those which correctly explained the variation in the plane of the ecliptic, the subject has become so stereotyped that there really seems nothing impossible to many writers in stating, that the pole of the ecliptic is the centre of the circle traced by the pole of the heavens; that the two never vary their distance, although there is a constant change in their distance of 45" per century; for this, in other words, is what is said when we admit a decrease of 45" per century in the obliquity.

It might be interesting to inquire what opinion would now be expressed relative to these two inconsistent movements, supposing they were put forward as novelties at the present day, and as sound and geometrically possible, and that for two hundred years it had been agreed that the centre of the circle traced by the pole was 6° from the pole of the ecliptic.

Had this latter movement been time-honoured, it

is probable that it would be stated that this one movement explained all the facts of the precession—the difference in the length of the years, the decrease in the obliquity of the ecliptic, the climatic changes shown by geology to have formerly occurred—and that it was actually a geometrical fact, not a theory, as the movement was now going on.

Any person who brought forward some vague movement of a circle which varied its distance from its centre, of a course being traced by the earth's axis which 'for a little while was a circle, but for a long time is not a circle,' would receive somewhat rough handling at the hands of geometricians; whilst practical men would probably state that such a movement for the earth's axis was a geometrical impossibility, and only explained partially what the other explained fully.

If a theorist suggested that the earth's axis traced a circle round the pole of the ecliptic as a centre, and never varied its distance from that centre, and this happened to be now a new theory, he would probably be advised to look-up his definitions in his Euclid, and note how a variable obliquity was in fact a variable radius. Also, probably this theory would not be considered worth much if its proposer could not tell what the curve traced really was, nor where the centre was, and yet assumed, in opposition to the fact of a decreasing obliquity, that the pole of the ecliptic must be the centre of this circle.

When an investigation of the course traced by

the pole of the heavens had demonstrated that this course was such as to bring the two poles to a distance from each other of 35° 25′ 47″, we found upon working out the climatic conditions that would occur, a great annual alternation of heat and cold in northern and southern latitudes, from 50° up to the poles.

Upon reflecting on these changes, we shortly recognised the fact that the heat of the summer must liberate an immense number of icebergs from the northern regions, and that unless there had been great heat, no icebergs could be liberated. When, however, we spoke to some of our geological friends, we found that they were under the impression that the glacial period must have been one long epoch of continued cold alone. Upon pressing the question, as to how the icebergs in high northern regions came to be freed formerly, or even now, we believe many of these gentlemen entirely altered their views, and came to the conclusion that heat must somehow have had a share in the effects of the glacial period; but how, or in what manner, they did not guess.

Having from a long examination of the effects of the boulder and drift period become quite convinced of the necessity of the heat, as well as of the cold, to produce the results, we were gratified to find that several of the most careful observers of the day are of the same opinion. Thus Professor Tyndall states that the ancient glaciers indicate the action of heat as much as they do cold. 'Cold,' he says, 'will not produce glaciers. You may have the bitterest north-east winds here in London throughout the winter without a single flake of snow. Cold must have the fitting object to operate upon, and this object—the aqueous vapour of air—is the direct product of heat. . . . It is perfectly manifest that by weakening the sun's action, either through a defect of emission or by the steeping of the entire solar system in space of a low temperature, we should be cutting-off the glaciers at their source.'

Speaking on this subject, Sir John Lubbock, in *Prehistoric Times*, page 297, says: 'To produce snow requires both heat and cold; the first to cause evaporation, the second to produce condensation. In fact, what we require is a greater contrast between the temperature of the tropics and that of our latitudes; so that, paradoxical as it may appear, the primary cause of the glacial epoch may be, after all, an elevation of temperature in the tropics, causing a greater amount of evaporation in the equatorial regions, and consequently a greater supply of the raw material of snow in the temperate regions during winter months.'

When we see that such conclusions have been arrived at by those thinkers who have studied the meteorological conditions requisite to produce the immense masses of snow which it appears covered the earth during the glacial epoch, and find that the course we have traced as that which the pole follows, gives exactly these conditions—the great heat in summer, the cold in winter, the great contrast between the uniform heat of the tropics and the arctic climate down to lat. 54°—we see no reason to be displeased with our geometrical problem, which during ten years has grown under our care, and from its first crude and imperfect state has developed powers which are to some extent shown in even this imperfect sketch.

When we dwell on the actual results due to an obliquity of 35° 25′, we cannot avoid realising the fact, that during the winters under such conditions the accumulation of snow and ice would even in latitudes of 54° be immense. Then the rapidly increasing north declination of the sun would cause a rapid thaw of all these masses of snow, and the denudations would also have been immense. The rush of waters produced by this melted snow would follow the natural water-course of the country, and such is found to be the case. In many instances where the valleys have little or no outlet, the water produced by the melted masses would there lie and deposit its various materials, forming vast beds of sand, clay, or gravel. Then the immense heat of the summer sun in this same latitude, the sun having a meridian altitude 12° more than now, and being twenty-four hours above the horizon, would produce enormous evaporation, which would probably leave the surface dry.

This alternation of vast quantities of water with great heat producing evaporation during about 16,000 years, viz. during the interval between the pole moving over 180° of its course, would admit of masses of sand, gravel, and clay being quietly deposited over miles of country, leaving the clay in the valleys, whilst the rolled pebbles were on the hill-sides; and all these effects produced without any grand catastrophes, vast upheaving waves, or any of those cataclysms, supposed by some writers to be the only means of explaining the facts.

During many years we have wandered in various countries where there was evidence of the glacial or drift period. England, Ireland, Scotland, France, and Brittany have all been examined. As we noted the evidence on the earth's surface, and pictured to ourselves the conditions resulting from what we knew in connection with the inclination of the earth's axis at a date 15,000 years in the past, it seemed as if a page of the world's history were lying open before us, on which were the words and sentences, telling us of what had occurred in the summer and what in the winter at that remote date. There we saw a vast boulder, on which were the markings and scratchings as plainly to be seen as when, 15,000 years ago, an iceberg bore it from those distant rocky hills and grounded it on the bank where it now rests. Here in this valley are the thin layers of gravel; each line is a summer's deposit, and these are now accountable. Resting on some giant and half-sunken boulders are several of smaller size: these were brought down when the pole was nearing that part of its circle 90° from where it now is, for then the obliquity would have been but 30°, and the date of large icebergs was past.

About 8000 B.C. would have been the period when these travellers came and were stranded on the spot where they have ever since remained, the hand of time scarcely leaving a mark on them, and they looking as if it were only yesterday they had come to their present resting-places.

We trust that after the geometrical proof we have given of the course traced by the earth's axis, some readers at least will pardon us for believing in the truth of this movement, as we have interpreted it; and those who do—after a searching examination—believe that it is *the* true explanation of the great glacial epoch, will realise the singular interest with which we examined all the effects of this movement, when we state that for above ten years we have worked out the principal facts connected with this problem, and have been able, without the aid of any observation, but by *a geometrical proof*, to trace out not only the present course of the pole for many years, but to arrive at the date at which this course must produce the conditions written so plainly on the ground around us.

The most extensive research has resulted in our failing to find any previous suggestion even of this course of the pole. The alteration in the earth's axis has been frequently suggested; but that is a very different thing from that which we have brought forward. We have evidence that for 300 years at least, and probably for 1800, there has been no alteration in the earth's axis; but we have direct evidence that for 1800 years this axis has changed its direction as regards the heavens, and during all those centuries it has steadily marked out a portion of that same circle, which 15,000 years ago, caused an obliquity of the ecliptic sufficient to produce the glacial epoch.

Thus we believe we may claim for this problem one peculiarity, viz. its novelty. Upwards of thirteen years ago we first submitted it in a very crude state to criticism in a public lecture;* and two years ago we gave the formula which is herein given for finding the obliquity. As, however, we did not then deem it necessary to supply any portion of the other evidence connected with the problem, the announcement caused much the same effect as was produced on Herodotus,† when he heard men state the world was round.

In the preceding pages, however, we believe those readers who are interested in investigations will find that they have a sufficiently distinct problem to look

---

* At the Leeds Philosophical Society, 1859.

† From Herodotus, *Melpomene*, xxxvi.: 'For my own part I cannot but think it exceedingly ridiculous to hear some men talk of the circumference of the earth, pretending, without the smallest reason or probability, that the ocean encompasses the earth, that the earth is round, as if mechanically formed so, and that Asia is equal to Europe. I will therefore concisely describe the figure and the size of each of these portions of the earth.'

The Rev. W. Beloe, in his foot-note to the above, remarks: 'We might be induced to conclude, from this incidental sneer of Herodotus, that there were some excellent astronomers and geographers in his time; although, like Copernicus and others, they did not obtain much credit among their contemporaries.'

into, and probably they may go over the same ground which we have so often trodden during the past few years. On comparing the various theories or conjectures relative to the cause of the glacial epoch, with the explanation afforded by the movement of the pole in a larger circle than has hitherto been assumed, we find the marked difference, that nearly every one of the former theories were conjectures, not based upon changes now in operation. We have, however, in operation the polar movement, and its course is so distinctly marked, that we have merely to continue this course in order to arrive at those very conditions requisite to produce the effects of the glacial epoch.

We, however, wish to point out, that the glacial epoch is merely *a corroboration of this movement, not the principal evidence of its occurrence.* That the corroboration is very strong will, we believe, be admitted; but the evidence of its occurrence is the geometrical accuracy with which results are obtained from this movement, and which are unattainable when we assume the impossible one of a centre changing its distance from a circumference.

If, then, it be urged that this movement of the pole is not the true one, we are warranted in inquiring from such objectors, how it is that the delicate formula which we have given, and which is based entirely on the correct localisation of the centre of this circle, gives results agreeing to a second with the best modern observations?

If this be not the true course of the pole, then what is its true course?

Are we still to state that it is a circle the centre of which is the pole of the ecliptic, from which centre it never varies its distance of $23° 28'$; and then immediately afterwards admit that it is always varying?

The movement of the pole as at present accepted contains so great a geometrical error, that it ought no longer to be admitted in works even of popular astronomy which are otherwise worthy of every attention. This movement is, as at present accepted, producing confusion in many most important astronomical problems, and it fails to afford any explanation of the facts which have so long and so carefully been collected by geologists, instead of yielding a solution of these as complete as can be required.

In the full confidence that a fair examination only is all that is required to show the importance of the problem which we have here described, we submit this work to public opinion and investigation.

In our next work the details of this problem will be investigated, and the results thereof as affecting changes in the right ascensions and declinations of celestial bodies will be demonstrated, as also will other results not hitherto known.

THE END.

Chapman and Hall's
CATALOGUE OF BOOKS;

INCLUDING

BOOKS FOR THE USE OF SCHOOLS,

ISSUED UNDER THE AUTHORITY OF

The Science and Art Department, South Kensington.

193, PICCADILLY, LONDON,
February, 1872.

THOMAS CARLYLE'S WORKS.

PEOPLE'S EDITION.

IN MONTHLY TWO-SHILLING VOLUMES,
Small crown 8vo.

With Mr. CARLYLE'S consent, we are issuing a Cheap Edition of his Works, printed from the Text of the LIBRARY EDITION, which is just completed, and which has been revised by himself.

The Volumes are handsomely printed in clear type, with good paper and cloth binding, and will consist of the following works:

SARTOR RESARTUS. 1 vol., with Portrait of Mr. Carlyle.
THE FRENCH REVOLUTION. 3 vols.
LIFE OF JOHN STERLING. 1 vol.
OLIVER CROMWELL'S LETTERS AND SPEECHES. 5 vols.
HERO-WORSHIP. 1 vol.
PAST AND PRESENT. 1 vol. [*In the Press.*
CRITICAL AND MISCELLANEOUS ESSAYS. 7 vols. [*In the Press.*

CHARLES DICKENS'S WORKS.

HOUSEHOLD EDITION.

IN WEEKLY PENNY NUMBERS,
AND SIXPENNY MONTHLY PARTS.
Each Penny Number will contain Two Illustrations.

The Series commenced on July 1st, 1871.

OLIVER TWIST. Complete in One Volume, with 28 Illustrations, cloth gilt, 2s. 6d.; in paper covers, 1s. 6d.
MARTIN CHUZZLEWIT. Complete in One Volume, with 59 Illustrations, cloth gilt, 4s.; in paper covers, 3s. [*February 24th.*

Messrs. CHAPMAN & HALL trust that by this Edition they will be enabled to place the Works of the most popular British Author of the present day in the hands of all English readers.

DAVID COPPERFIELD *will be the next work issued.*

New Publications.

The Tenth Edition is now ready.

THE LIFE OF CHARLES DICKENS.
By JOHN FORSTER.
Vol. I. 1812—1842.
With Portraits and other Illustrations. Demy 8vo. Price 12s.

In Two Handsome Volumes. Price £1 4s.

THE KERAMIC GALLERY,
Comprising about Six Hundred Illustrations of rare, curious, and choice examples of Pottery and Porcelain, from the Earliest Times to the Present, selected by the Author from the British Museum, the South Kensington Museum, the Geological Museum, and various Private Collections. With Historical Notices and Descriptions.
By WILLIAM CHAFFERS,
Author of "Marks and Monograms on Pottery and Porcelain," "Hall Marks on Plate," &c.

THE HIGHLANDS OF CENTRAL INDIA.
NOTES ON THEIR FORESTS, WILD TRIBES, NATURAL HISTORY, AND SPORTS.
By THE LATE CAPT. J. FORSYTH.
With a Map and Coloured Illustrations. Demy 8vo. Price 14s.

VOLTAIRE.
By JOHN MORLEY.
Demy 8vo. Price 14s.

THE LIFE OF OLIVER GOLDSMITH.
By JOHN FORSTER.
Fifth Edition. With additional Notes, the original Illustrations by MACLISE, STANFIELD, LEECH, DOYLE, several additional designs, and two beautifully engraved Portraits from the Original Painting by REYNOLDS and from the Statue by FOLEY. In 2 vols. Price 21s.

SIR JOHN ELIOT:
A BIOGRAPHY.
By JOHN FORSTER.
A New and Popular Edition, with Portraits. In 2 Vols. Price 14s.

WALTER SAVAGE LANDOR:
A BIOGRAPHY.
By JOHN FORSTER.
New and Cheaper Edition, with Portraits. In 1 Vol.   [*In the Press.*

MR. THOMAS CARLYLE'S WORKS.
THE LIBRARY EDITION IS NOW COMPLETE IN THIRTY-THREE VOLUMES.
Demy 8vo, with Portraits and Maps.
A GENERAL INDEX TO THE ABOVE. In One Vol., demy 8vo. Price 6s.

THE DIARY OF AN IDLE WOMAN IN ITALY.
By FRANCES ELLIOT.
New Edition, in 1 Vol. Price 6s.

PICTURES OF OLD ROME.
By MRS. ELLIOT.
New Edition, in 1 Vol. Price 6s.

LETTERS AND EXTRACTS FROM THE
OCCASIONAL WRITINGS OF J. BEETE JUKES,
M.A., F.R.S., late Director of the Geological Survey in Ireland.
EDITED, WITH MEMORIAL NOTES, BY HIS SISTER.
Comprising Letters from Australia and Newfoundland whilst engaged in the Geological Survey.
Post 8vo., with a Portrait. Price 12s.

HISTORY OF ENGLAND FROM THE YEAR 1830.
By WILLIAM NASSAU MOLESWORTH.
Vol. I., demy 8vo. Price 15s. [Vol. II. in the Press.

THE SECOND SERIES OF
THE EARTH.
A DESCRIPTIVE HISTORY OF THE PHENOMENA AND LIFE OF THE GLOBE.
THE SECOND SERIES.
THE OCEAN, THE ATMOSPHERE, AND LIFE.
By ÉLISÉE RECLUS.
With 207 Illustrations and 27 coloured Maps. Demy 8vo. [In the Press.

THE FAMILY MEDICAL GUIDE.
WITH PLAIN DIRECTIONS FOR THE TREATMENT OF EVERY CASE, AND A LIST OF MEDICINES REQUIRED FOR ANY HOUSEHOLD.
By GEORGE FULLERTON, C.M. and M.D., EDINBURGH.
Demy 8vo. Price 12s.

THE STORY OF THE COMMUNE.
By A COMMUNALIST.
Reprinted from the "Pall Mall Gazette," with Additions. 8vo, price 1s.

THE HISTORY OF THE COMMUNE OF PARIS.
By P. VESINIER,
Ex-member and Secretary of the Commune, and Rédacteur en Chef du *Journal Officiel*.
One Vol., crown 8vo. Price 7s. 6d.

BOOKS PUBLISHED BY

THE EARTH.

A DESCRIPTIVE HISTORY OF THE PHENOMENA AND LIFE OF THE GLOBE.

By ÉLISÉE RECLUS.

Translated by the late B. R. WOODWARD, and Edited by HENRY WOODWARD.

With 234 Maps and Illustrations, and 24 page Maps printed in colours.

2 vols. large demy 8vo, 26s.

ANIMAL PLAGUES: THEIR HISTORY, NATURE, AND PREVENTION.

By G. FLEMING, R.E., F.R.G.S.,

Author of "Horse-Shoes and Horse Shoeing."

Demy 8vo, 15s.

ENGLISH PREMIERS

FROM SIR ROBERT WALPOLE TO SIR ROBERT PEEL.

By J. C. EARLE.

2 vols. post 8vo, 21s.

CRITICAL AND MISCELLANEOUS ESSAYS.

By JOHN MORLEY.

Demy 8vo, 14s.

EXPERIENCES OF A PLANTER IN THE JUNGLES OF MYSORE.

By ROBERT H. ELLIOT.

2 vols. demy 8vo, with a Map and Illustrations, 24s.

CLEMENT MAROT, AND OTHER STUDIES.
By HENRY MORLEY.
In 2 vols., crown 8vo, 18s.

RODA DI ROMA.
By W. W. STORY.
Sixth Edition, In one Volume, with a Portrait, 10s. 6d.

MODERN WAR,
OR THE CAMPAIGNS OF THE FIRST PRUSSIAN ARMY DURING THE WAR 1870-71.
By SIR RANDAL ROBERTS, BART.
8vo, cloth, with a Map, 14s.

PARIS DURING THE SIEGE.
By FRANCISQUE SARCEY.
Post 8vo, cloth, 6s. 6d.

THE GOLDEN AGE, A SATIRE.
By ALFRED AUSTIN.
Post 8vo, cloth, 7s.

THE OLD COLONEL AND THE OLD CORPS
WITH A VIEW OF MILITARY ESTATES.
By LIEUT.-COL. G. R. S. GLEIG.
Post 8vo, cloth, 8s.

New Novels.

CASTAWAY. By EDMUND YATES. 3 vols. *In February.*
BROKEN TOYS. A NOVEL. By ANNA STEELE. 3 vols. *In February.*
MEN WERE DECEIVERS EVER. A HISTORY. By HAMILTON MARSHALL. 2 vols. *In February.*
FINGER OF FATE. A ROMANCE. By CAPTAIN MAYNE REID. 2 vols. *In February.*
ETHEL MILDMAY'S FOLLIES. A STORY. By the Author of "Petite's Romance." 3 vols. *In the Press.*
ONLY THREE WEEKS. A NOVEL. By the Author of "Ereighda Castle." 3 vols.
THE VALLEY OF POPPIES. By JOSEPH HATTON. 2 vols.
THE ROSE AND THE KEY. By J. SHERIDAN LE FANU. 3 vols.
KIMBERWELL HOUSE. By ROBERT HUDSON. 3 vols.
HELEN CAMERON; FROM GRUB TO BUTTERFLY. By the Author of "Mary Stanley." 3 vols.
MAGDALEN WYNWARD; OR THE PROVOCATIONS OF A PRE-RAPHAELITE. By AVERIL BEAUMONT. 2 vols.
A TERRIBLE TEMPTATION. A STORY OF THE DAY. By CHARLES READE. 3 vols.
FOLLE-FARINE. A NEW NOVEL. by "OUIDA." 3 vols.
SARCHEDON; A LEGEND OF THE GREAT QUEEN. By G. J. WHYTE-MELVILLE. 3 vols.

WHYTE-MELVILLE'S WORKS.

Re-issue in Two-Shilling Vols., fancy boards, or 2s. 6d. in cloth.

THE WHITE ROSE.
CERISE. A Tale of the Last Century.
THE BROOKES OF BRIDLEMERE.
"BONES AND I;" or, The Skeleton at Home.
SONGS AND VERSES.
MARKET HARBOROUGH; or, How Mr. Sawyer went to the Shires.
CONTRABAND; or, a Losing Hazard.
M. OR N.—Similia Similibus Curantur.

BOOKS

PUBLISHED BY

Chapman and Hall.

Abd-el-Kader: a Biography. Written from dictation by COLONEL CHURCHILL. With fac-simile letter. Post 8vo, 9s.

All the Year Round. Conducted by Charles Dickens. First Series. 20 vols. royal 8vo, cloth, 5s. 6d. each.

———— New Series. Vols. 1 to 6, royal 8vo, cloth, 5s. 6d. each.

———— The Christmas Numbers, in 1 vol. Boards, 2s. 6d.; cloth, 3s. 6d.

Army Misrule. By a Common Soldier. Second Edition. Post 8vo, cloth, 3s. 6d.

Austro-Hungarian Empire and the Policy of Count Beust. A Political Sketch of Men and Events from 1866 to 1870. By an ENGLISHMAN. Second Edition. Demy 8vo, with Maps. 9s.

    Part I.—The New Constitution.
   ,, II.—Foreign Policy.
   ,, III.—Question of the Nationalities, Electoral Reform, the late Ministerial Crisis.

Bell (Dr. W. A.)—New Tracks in North America. A Journal of Travel and Adventure, whilst engaged in the Survey of a Southern Railroad to the Pacific Ocean, during 1867—8. With twenty Chromos and numerous Woodcuts. Second edition, demy 8vo, 18s.

Benson's (W.)—Principles of the Science of Colour. Small 4to, cloth, 15s.

———— **Manual of the Science of Colour.** *Coloured Frontispiece and Illustrations.* 12mo, cloth, 2s. 6d.

Blyth (Colonel)—The Whist-Player. With Coloured Plates of "Hands." Third edition, imp. 16mo, cloth, 5s.

Bolton (M. P. W.)—Inquisitio Philosophica; an Examination of the Principles of Kant and Hamilton. *New Edition.* Demy 8vo, cloth. 8s. 6d.

——— Examination of the Principles of the Scoto-Oxonian Philosophy. *New Edition.* Demy 8vo, cloth, 5s.

Bowden (Rev. J.)—Norway, its People, Products, and Institutions. Crown 8vo, 7s. 6d.

Brackenbury (Captain, C.B.)—Foreign Armies and Home Reserves. Republished by special permission from the *Times*. Crown 8vo, cloth, 5s.

——— European Armaments in 1867. Post 8vo, 5s.

——— The Constitutional Forces of Great Britain. A Lecture. Crown 8vo, sewed, 1s.

Bradley (Thomas), of the Royal Military Academy, Woolwich—Elements of Geometrical Drawing. In two Parts, with Sixty Plates, oblong folio, half bound, each part, 16s.

——— Selection (from the above) of Twenty Plates, for the use of the Royal Military Academy, Woolwich. Oblong folio, half bound, 10s.

Brookes of Bridlemere. By WHYTE MELVILLE.

Buchanan (Robert)—The Land of Lorne; including the Cruise of "The Tern" to the Outer Hebrides. 2 vols. post 8vo, cloth, 21s.

Buckmaster (J. C.)—The Elements of Mechanical Physics. With numerous Illustrations, fcap. 8vo, cloth, 3s.

Burchett (R.)—Linear Perspective, for the Use of Schools of Art. Fourteenth edition, with Illustrations, post 8vo, cloth, 7s.

——— Practical Geometry, the Course of Construction of Plane Geometrical Figures, with 137 Diagrams. Twelfth edition, post 8vo, cloth, 5s.

——— Definitions of Geometry. New edition, 24mo, cloth, 5d.

Carlyle (Dr.)—Dante's Divine Comedy.—Literal Prose Translation of THE INFERNO, with Text and Notes. Post 8vo. Second edition, 14s.

Carlyle (Thomas), Passages selected from his Writings. With Memoir. By THOMAS BALLANTYNE. Second edition, crown 8vo, 6s.

——— Shooting Niagara: and After? Crown 8vo. sewed, 6d.

——— Inaugural Address at Edinburgh, April 2, 1866, on being installed as Rector of the University there. By Thomas Carlyle. Sewed, 1s.

THOMAS CARLYLE'S WORKS.
LIBRARY EDITION COMPLETE.
HANDSOMELY PRINTED IN DEMY 8VO, CLOTH.

Sartor Resartus. The Life and Opinions of Herr Teufelsdrockh. With a Portrait, 7s. 6d.
The French Revolution. A History. 3 vols., each 9s.
Life of Frederick Schiller and Examination of His Writings. With Portrait and Plates, 7s. 6d.
Critical and Miscellaneous Essays. 6 vols., each 9s.
On Heroes, Hero Worship, and the Heroic in History. With a Portrait, 7s. 6d.
Past and Present. With a Portrait, 9s.
Oliver Cromwell's Letters and Speeches. With Portraits, 5 vols., each 9s.
Latter-Day Pamphlets. 9s.
Life of John Sterling. With Portrait, 9s.
History of Frederick the Second. 10 vols., each 9s.
Translations from the German. 3 vols., each 9s.
General Index to the Library Edition. 8vo, cloth.

THOMAS CARLYLE'S WORKS.
CHEAP AND UNIFORM EDITION.
In crown 8vo, cloth.

The French Revolution: A History. In 2 vols., 12s.
Oliver Cromwell's Letters and Speeches, with Elucidations, &c. 3 vols., 18s.
Life of Schiller } 1 vol., 6s.
Life of John Sterling
Critical and Miscellaneous Essays. 4 vols., 1l. 4s.
Sartor Resartus }
Lectures on Heroes } 1 vol., 6s.
Latter-Day Pamphlets, 1 vol., 6s.
Chartism }
Past and Present } 1 vol., 6s.
Translations from the German of Musæus, Tieck, and Richter. 1 vol., 6s.
Wilhelm Meister, by Goethe, a Translation, 2 vols., 12s.
History of Friedrich the Second, called Frederick the Great.
Vols. I. and II., containing Part I.—"Friedrich till his Accession." 14s.
Vols. III. and IV., containing Part II.—"The First Two Silesian Wars, and their result." 14s.
Vols. V., VI., VII., completing the Work, 1l. 1s.

THOMAS CARLYLE'S WORKS.
PEOPLE'S EDITION.
Volumes already published.

IN SMALL CROWN 8VO. PRICE 2s. EACH VOL. BOUND IN CLOTH.

Sartor Resartus. 2s.
French Revolution. 3 Vols., 6s.
Life of John Sterling. 2s.
Oliver Cromwell's Letters and Speeches. 5 Vols. 2s. each.
On Heroes and Hero Worship. 2s.
Past and Present. 2s.

Cecil Castlemaine's Gage, and other Novelettes.—By OUIDA.
With Frontispiece, crown 8vo, 5s.

Chandos. By OUIDA. With Frontispiece, crown 8vo, 5s.

Chronicles and Characters. By the HON. ROBERT LYTTON (OWEN MEREDITH). 2 vols., crown 8vo. Portrait, 1l. 4s.

Craik (George Lillie)—English of Shakespeare. Illustrated in a Philological Commentary on his Julius Cæsar. Fourth edition, post 8vo, cloth, 5s.

—— **Outlines of the History of the English Language.** Eighth edition, post 8vo, cloth, 2s. 6d.

Dante.—Dr. J. A. Carlyle's Literal Prose Translation of the Inferno, with the Text and Notes. Second edition. Post 8vo, 14s.

D'Aumale (Le Duc)—The Military Institutions of France, by H.R.H. The Duc D'AUMALE. Translated with the Author's consent by Capt. Ashe, King's Dragoon Guards. Post 8vo, 6s.

D'Azeglio—Recollections of the Life of Massimo D'Azeglio. Translated, with an Introduction and Notes, by COUNT MAFFEI. 2 vols., post 8vo, 1l. 4s.

De Coin (Colonel Robert L.)—History and Cultivation of Cotton and Tobacco. Post 8vo, cloth, 9s.

De la Chapelle (Count).—The War of 1870. Events and Incidents of the Battle Field. Post 8vo, cloth, 4s. 6d.

De Guérin (Maurice and Eugénie). A Monograph. By HARRIETT PARR, Author of "Essays in the Silver Age," &c., crown 8vo, cloth, 6s.

De Leon (Edwin).—Askaros Kassis, the Copt, a Romance of Modern Egypt. Crown 8vo, cloth, 7s. 6d.

De Quetteville (Rev. P. W.).—The Pardon of Guingamp, or Poetry and Romance in Modern Brittany. Post 8vo, cloth, 9s.

Dinners and Dinner Parties; or, The Absurdities of Artificial Life, with Additions, including a short Catechism of Cookery, founded on the principles of Chemistry. Second edition, post 8vo, cloth, 3s. 6d.

Dixon (W. Hepworth)—The Holy Land. Fourth Edition, with 2 Steel and 12 Wood Engravings, post 8vo, 10s. 6d.

—— **William Penn:** An Historical Biography, founded on Family and State Papers. With a New Preface in reply to the accusations of Mr. Macaulay. Small crown 8vo. Portrait, 7s.

—— **Robert Blake**, Admiral and General at Sea. Based on Family and State Papers. Crown 8vo. Portrait. 2s. 6d.

CHARLES DICKENS'S WORKS.

ORIGINAL EDITIONS.

The Mystery of Edwin Drood. With Illustrations by S. L. Fildes, and a Portrait engraved by Baker. 8vo, 7s. 6d. cloth.

Our Mutual Friend. With Forty Illustrations by Marcus Stone. Demy 8vo, cloth, 1l. 1s.

The Pickwick Papers. With Forty-three Illustrations by Seymour and 'Phiz.' Demy 8vo, cloth, 1l. 1s.

Nicholas Nickleby. With Forty Illustrations by 'Phiz.' Demy 8vo, cloth, 1l. 1s.

Sketches by 'Boz.' With Forty Illustrations by George Cruikshank. Demy 8vo, cloth, 1l. 1s.

Martin Chuzzlewit. With Forty Illustrations by 'Phiz.' Demy 8vo, cloth, 1l. 1s.

Dombey and Son. With Forty Illustrations by 'Phiz.' Demy 8vo, cloth, 1l. 1s.

David Copperfield. With Forty Illustrations by 'Phiz.' Demy 8vo, cloth, 1l. 1s.

Bleak House. With Forty Illustrations by 'Phiz.' Demy 8vo, cloth, 1l. 1s.

Little Dorrit. With Forty Illustrations by 'Phiz.' Demy 8vo, cloth, 1l. 1s.

Oliver Twist and Tale of Two Cities. In One Volume. Demy 8vo, cloth, 21s.

The Old Curiosity Shop. With Seventy-five Illustrations by George Cattermole and H. K. Browne. A New Edition, Demy 8vo, uniform with the other Volumes.

Barnaby Rudge: a Tale of the Riots of 'Eighty. With Seventy-eight Illustrations by G. Cattermole and H. K. Browne. Demy 8vo, uniform with the other Volumes.

Oliver Twist. With Twenty-four Illustrations. Demy 8vo, cloth, 11s.

A Tale of Two Cities. With Sixteen Illustrations by 'Phiz.' Demy 8vo, cloth, 9s.

Hard Times. Small 8vo, cloth, 5s.

The Uncommercial Traveller. Crown 8vo, cloth, 6s.

CHARLES DICKENS'S WORKS.

ILLUSTRATED LIBRARY EDITION.

With the Original Illustrations, 26 vols., post 8vo, cloth, 8s. per volume. £ s. d.

Pickwick Papers	With 43 Illustrations,	2 vols.	...	0 16 0
Nicholas Nickleby	With 39	2 vols.	...	0 16 0
Martin Chuzzlewit	With 40	2 vols.	...	0 16 0
Old Curiosity Shop and Reprinted Pieces	With 36	2 vols.	...	0 16 0
Barnaby Rudge & Hard Times	With 36	2 vols.	...	0 16 0
Bleak House	With 40	2 vols.	...	0 16 0
Little Dorrit	With 40	2 vols.	...	0 16 0
Dombey and Son	With 38	2 vols.	...	0 16 0
David Copperfield	With 38	2 vols.	...	0 16 0
Our Mutual Friend	With 40	2 vols.	...	0 16 0
Sketches by Boz	With 39	1 vol.	...	0 8 0
Oliver Twist	With 24	1 vol.	...	0 8 0
Christmas Books	With 17	1 vol.	...	0 8 0
A Tale of Two Cities	With 16	1 vol.	...	0 8 0
Great Expectations	With 8	1 vol.	...	0 8 0
Pictures from Italy, and American Notes	With 8	1 vol.	...	0 8 0

THE "CHARLES DICKENS" EDITION.

In 10 vols. Crown 8vo, with Illustrations.

Pickwick Papers	With 8 Illustrations	...	0 3 6
Martin Chuzzlewit	With 8	,,	...	0 3 6
Dombey and Son	With 8	,,	...	0 3 6
Nicholas Nickleby	With 8	,,	...	0 3 6
David Copperfield	With 8	,,	...	0 3 6
Bleak House	With 8	,,	...	0 3 6
Little Dorrit	With 8	,,	...	0 3 6
Our Mutual Friend	With 8	,,	...	0 3 6
Tales of Two Cities	With 8	,,	...	0 3 0
Sketches by Boz	With 8	,,	...	0 3 0
American Notes, and Reprinted Pieces	With 8	,,	...	0 3 0
Barnaby Rudge	With 8	,,	...	0 3 0
Christmas Books	With 8	,,	...	0 3 0
Old Curiosity Shop	With 8	,,	...	0 3 0
Oliver Twist	With 8	,,	...	0 3 0
Great Expectations	With 8	,,	...	0 3 0
Hard Times, and Pictures from Italy	With 8	,,	...	0 3 0
Uncommercial Traveller	With 4	,,	...	0 3 0
A Child's History of England	With 4	,,	...	0 3 6

The above Works are sold separately.

DICKENS—THE LIFE OF CHARLES DICKENS.

By JOHN FORSTER. Vol. I., 1812-42, *with Portraits and other Illustrations*. 10th Edition. 8vo, cloth, 12s.

CHARLES DICKENS'S WORKS.
HOUSEHOLD EDITION,
Now in course of publication in Weekly Numbers at One Penny, and in Monthly Parts at Sixpence.

Each penny number contains two new illustrations.

OLIVER TWIST, with 28 Illustrations. Complete crown 4to, sewed, 1s. 6d.; in cloth, 2s. 6d.

MARTIN CHUZZLEWIT, with 59 Illustrations. Will be published on February 26th. Sewed, 3s.; in cloth, 4s.

MR. DICKENS'S READINGS.
Fcap. 8vo, sewed.

	£	s.	d.
Christmas Carol in Prose	0	1	0
Cricket on the Hearth	0	1	0
Chimes: A Goblin Story	0	1	0
Story of Little Dombey	0	1	0
Poor Traveller, Boots at the Holly-Tree Inn, and Mrs. Gamp	0	1	0

The Christmas Books: in one Volume, containing—The Christmas Carol; The Cricket on the Hearth; The Chimes; The Battle of Life; The Haunted House. *With all the original Illustrations. A handsome Volume.* Demy 8vo, cloth gilt 0 12 0

Dottings on the Roadside in Panama, Nicaragua, and Mosquito, by Captain Bedford Pim, R.N., and Berthold Seeman, Ph.D., F.L.S., F.R.G.S. Illustrated with Plates and Maps. Demy 8vo, 18s.

Dramatists of the Present Day. By Q. *Reprinted from the "Athenæum."* Post 8vo, cloth, 4s.

Drayson (Lieut.-Col. A. W.)—Practical Military Surveying and Sketching. 3rd Edition. Post 8vo, cloth, 4s. 6d.

Dyce's Shakespeare. New Edition, in Nine Volumes, demy 8vo.
—The Works of Shakespeare. Edited by the Rev. Alexander Dyce. This edition is not a mere reprint of that which appeared in 1857, but presents a text very materially altered and amended from beginning to end, with a large body of critical Notes almost entirely new, and a Glossary, in which the language of the poet, his allusions to customs, &c., are fully explained. 9 vols., demy 8vo, 4l. 4s.

"The best text of Shakespeare which has yet appeared.........Mr. Dyce's Edition is a great work, worthy of his reputation, and for the present it contains the standard text."—*Times.*

Dyce (Rev. A.)—A Glossary to the Works of Shakespeare. Demy 8vo, 12s.

⁎⁎ This forms Vol. 9 of the Rev. A. Dyce's edition of 'The Works of William Shakespeare,' and is sold separately.

Dyce (William), R.A.—Drawing-Book of the Government School of Design, or Elementary Outlines of Ornament. Fifty Selected Plates, folio, sewed, 5s.

———— The same, mounted, 18s.

———— Fifteen selected Outlines (from the above), mounted, 5s.

Earle's (J. C.) English Premiers, from Sir Robert Walpole to Sir Robert Peel. 2 vols. Post 8vo, cloth, 21s.

Eastwick (E. B.)—Venezuela: Sketches of Life in a South American Republic. With a Map. Second Edition. 8vo, cloth, 16s.

Egmont: a Tragedy. By Goethe. Translated from the Original German, by Arthur Duke Coleridge, M.A. With Entr'actes and Songs by Beethoven, newly arranged from the Full Score, and Schubert's Song, 'Freudvoll und Leidvoll,' and an Illustration by J. E. Millais, Esq., R.A. Crown 8vo, bevelled cloth, 8s. 6d.

Elementary Drawing-Book. Directions for Introducing the First Steps of Elementary Drawing in Schools and among Workmen. Small 4to, cloth, 4s. 6d.

Elementary Drawing Copy-Books, for the Use of Children from four years old and upwards, in Schools and Families. Compiled by a Student certificated by the Science and Art Department as AN ART TEACHER. Three Books in 4to, sewed :—

Book 1. LETTERS, 1s.
" 2. GEOMETRICAL AND ORNAMENTAL FORMS AND OBJECTS, 1s.
" 3. LEAVES, FLOWERS, SPRAYS, &c., 1s. 6d.

Elliot (Sir John)—A Biography by John Forster. With Portraits. A new and cheaper Edition. 2 vols. Post 8vo, cloth, 11s.

Elliot's (Robert H.) Experiences of a Planter in the Jungles of Mysore. With Illustrations and a Map. 2 vols, 8vo, cloth, 21s.

Elliot (Frances)—The Diary of an Idle Woman in Italy. 2nd Edition. Post 8vo, cloth, 6s.

—————— **Pictures of Old Rome.** New Edition. Post 8vo, cloth, 6s.

Finlaison (Alexander Glen)—New Government Succession-duty Tables. 3rd Edition. Post 8vo, cloth, 5s.

Fitzgerald (M. Purcell)—The Crowned Hippolytus of Euripides. with a selection from the Pastoral and Lyric Poets of Greece. Fcap. cloth, 7s.

Five Weeks in a Balloon. A Voyage of Exploration and Discovery in Central Africa. From the French of Jules Verne, with 64 Illustrations, post 8vo, 7s. 6d.

Fleming (Jas. M.)—Carmina Vitæ and other Poems. Post 8vo, cloth, 2s. 6d.

Fleming (George)— Animal Plagues, their History, Nature, and Prevention. 8vo, cloth, 15s.

—————— **Horses and Horse-shoeing;** their Origin, History, Uses and Abuses. 210 Engravings. 8vo, cloth, 1l. 1s.

Folle-Farine (a New Novel), by Ouida. 3 vols. Post 8vo, cloth, 1l. 11s. 6d.

Forest Life in Acadie.—Sketches of Sport and Natural History in the Lower Provinces of the Canadian Dominions, by Captain Campbell Hardy, R.A., with Illustrations. Demy 8vo, 18s.

Forster (John)—Oliver Goldsmith: a Biography. With Illustrations. In 2 vols. Large crown 8vo, 21s.

——— **Walter Savage Landor.** A Biography. 1775-1864. 2 vols. With Portraits and Vignettes. Post 8vo, 1l. 8s.

——— **Sir John Elliot.** A Biography. *With Portraits. New and cheaper Edition.* 2 vols. Post 8vo, cloth, 14s.

——— **Life of Charles Dickens.** Vol. I., 1812-42. With Portraits and other Illustrations. Tenth Edition, 8vo, cloth, 12s.

Forsyth (Capt.)—The Highlands of India. Notes on their Forests and Wild Tribes, Natural History, and Sports. With Map and Coloured Illustrations, 8vo, cloth, 18s.

Fortnightly Review.—First Series, May, 1865, to Dec. 1866. 6 vols., cloth.

——————————— New Series, 1867 to Present Time. In Half-yearly Volumes. Cloth, 13s. each; and Parts, 2s. each, *Monthly*.

Francatelli (C. E.)—Royal Confectioner. English and Foreign. A Practical Treatise. With Coloured Illustrations. New edition, post 8vo, cloth, 9s.

Fullerton (George)—Family Medical Guide. With plain Directions for the Treatment of every Case, and a List of Medicines required for any Household. 8vo, cloth, 12s.

Gentlewoman, The. By the Author of 'Dinners and Dinner Parties.' With Illustrations. Second Edition, post 8vo, cloth, 4s. 6d.

German Evenings. Tales. Translated from the Original by J. L. Lowdell. Crown 8vo, 7s. 6d.

Gillmore Parker ("Ubique")—All Round the World. Adventures in Europe, Asia, Africa, and America. With Illustrations by Sydney P. Hall. Post 8vo, cloth gilt, 7s. 6d.

Gleig's (Lt.-Col. C. S. E.)—The Old Colonel and the Old Corps; with a View of Military Estates. Post 8vo, cloth, 8s.

Hake (Thos. Gordon) — Madeline, with other Poems and Parables. Post 8vo, cloth, 7s. 6d.

Hall (Sidney)—A Travelling Atlas of the English Counties. Fifty Maps, coloured. New edition, including the railways, demy 8vo, in roan tuck, 10s. 6d.

Hardy (Captain Campbell)—Forest Life in Acadie. Sketches of Sport and Natural History in the Lower Provinces of the Canadian Dominion. With Plates and Coloured Frontispiece. Demy 8vo, 18s.

Hawkins (B. W.)—Comparative View of the Human and Animal Frame. Small folio, cloth, 12s.

Held in Bondage; or, Granville de Vigne. By OUIDA.
Post 8vo, cloth, 5s.

Holbein (Hans)—Life. By R. N. WORNUM. With Portrait and 31 Illustrations. Imp. 8vo, cloth, 31s. 6d.

Hulme (F. E.)—A Series of 60 Outline Examples of Free-hand Ornament. Royal 8vo, half roan, 10s. 6d.

Humphris (H. D.)—Principles of Perspective. Illustrated in a Series of Examples. Oblong folio, half bound, and Text 8vo, cloth, 21s.

Hutchings (James M.)—Scenes of Wonder and Curiosity in California. With above 100 Illustrations, demy 8vo, cloth, 12s.

Hutchinson (Alexander A., Capt. R.A.)— Try Lapland; a Fresh Field for Summer Tourists, with Illustrations and Map. Second edition, crown 8vo, cloth, 6s.

Jephson and Elmhirst.—Our Life in Japan. By R. MOUNTENEY JEPHSON, and E. PENNELL ELMHIRST, 9th Regt. With numerous Illustrations from Photographs by Lord WALTER KERR, SIGNOR BEATO, and native Japanese Drawings, 8vo, cloth, 18s.

Jukes (J. Beete).—Letters and Extracts from his Letters and Occasional Writings. Edited with Memorial Notes by his Sister. Portraits Post 8vo, cloth, 12s.

Kebbel (T. E.)—The Agricultural Labourer. A Short Survey of his Position. Crown 8vo, 6s.

Kenward (Jas.)—Oriel; a Study in 1870. With Two other Poems. Post 8vo, cloth, 5s.

Keramic Gallery. Comprising upwards of 500 Illustrations of rare, curious, and choice examples of Pottery and Porcelain, from the Earliest Times to the Present, selected by the Author from the British Museum, the South Kensington Museum, the Geological Museum, and various Private Collections. With Historical Notices and Descriptions. By WILLIAM CHAFFERS. Two handsome Vols. Price 4l. 4s.

Lacordaire (Rev. Père)—Jesus Christ. Conferences delivered at Notre Dame in Paris. Translated, with the author's permission, by a Tertiary of the same order. Crown 8vo, cloth, 6s.

——— **God.** Conferences delivered at Notre Dame, in Paris. By the same Translator. Crown 8vo, cloth, 6s.

Landor's (Walter Savage) Works. 2 vols., royal 8vo, cloth, 21s.

——————————— **A Biography.** 1775-1864. By JOHN FORSTER. Portraits and Vignettes. 2 vols., post 8vo, 1l. 8s.

Leroy (Charles Georges)—The Intelligence and Perfectibility of Animals, from a Philosophic Point of View, with a few Letters on Man. Post 8vo, cloth, 7s. 6d.

CHARLES LEVER'S WORKS.

Complete in 19 Vols., 12mo, Roxburgho binding, 3l. 3s.

RAILWAY EDITION.

12mo, fancy boards.

	s. d.		s. d.
Martins of Cro' Martin...	3 0	One of Them................	2 0
Charles O'Malley..........	3 0	Sir Jasper Carew............	2 0
The Daltons	3 0	Maurice Tiernay............	2 0
Davenport Dunn	3 0	A Day's Ride; A Life's Ro-	
The Knight of Gwynne ..	3 0	mance	2 0
Dodd Family	3 0	Jack Hinton	2 0
Roland Cashel	3 0	Barrington	3 0
Tom Burke	3 0	Luttrell of Arran	2 0
The O'Donoghue	2 0	St. Patrick's Eve, and Beat in	
Glencore	2 0	a Cloud	2 0
Harry Lorrequer............	2 0	Paul Gosslett's Confessions...	1 0

Levy's (W. Hanks) Blindness and the Blind; or a Treatise on the Science of Typhology. Post 8vo, cloth, 7s. 6d.

Lytton (Hon. Robt.)—' Owen Meredith.'—Orval; or the Fool of Time, and other Imitations and Paraphrases. 12mo, cloth, 9s.

—— **Chronicles and Characters.** With Portrait. 2 vols., crown 8vo, cloth, 1l. 4s.

—— **Collective Edition of "Owen Meredith's" Poetical Works—**

Vol. I.—CLYTEMNESTRA, and Poems Lyrical and Descriptive, 12mo, cloth, 6s.

,, II.—LUCILE. 12mo, cloth, 6s.

—— **Serbski Pesme;** or, National Songs of Servia. Fcap. cloth, 4s.

Lytton (Lord)—Money. A Comedy. Demy 8vo, sewed, 2s. 6d.

—— **Not so Bad as we Seem.** A Comedy. Demy 8vo, sewed, 2s. 6d.

—— **Richelieu; or, The Conspiracy.** A Play. Demy 8vo, sewed, 2s. 6d.

—— **Lady of Lyons, or Love and Pride.** A Play. Demy 8vo, sewed, 2s. 6d.

Mallet (Dr. J. W.)—Cotton: the Chemical, &c., Conditions of its Successful Cultivation. Post 8vo, cloth, 7s. 6d.

Mallet (Robert) — Great Neapolitan Earthquake of 1857. First Principles of Observational Seismology; as developed in the Report to the Royal Society of London, of the Expedition made into the Interior of the Kingdom of Naples, to Investigate the Circumstances of the great Earthquake of December, 1857. Maps and numerous Illustrations. 2 vols. royal 8vo, cloth, 63s.

Melville (G. J. Whyte) — Sarchedon, a Legend of the Great Queen. 3 Vols. Post 8vo, cloth. 1l. 11s. 6d.

CHEAP EDITION OF WHYTE-MELVILLE'S WORKS.

Crown 8vo, fancy boards, 2s. each, or 2s. 6d. in cloth.

The White Rose.

Cerise. A Tale of the Last Century.

Brookes of Bridlemere.

'Bones and I;' or, The Skeleton at Home.

Songs and Verses.

"M., or N." Similia Similibus Curantur.

Contraband, or a Losing Hazard.

Market Harborough; or, How Mr. Sawyer went to the Shires.

Memorials of Theophilus Trinal. By T. T. Lynch. New Edition, enlarged. Crown 8vo, cloth, extra, 6s.

Meredith (Owen). — See LYTTON, HON. ROBERT.

Meredith (George) — Shaving of Shagpat. An Arabian Entertainment. Post 8vo, cloth, 5s.

——— **Modern Love, and Poems of the English Roadside,** with Poems and Ballads. Fcap. cloth, 6s.

Milton's (John) Life, Opinions, and Writings. With an Introduction to "Paradise Lost," by THOMAS KEIGHTLEY. 8vo, cloth, 10s. 6d.

Molesworth (W. Nassau) — History of England from the Year 1830. Vol. 1. 8vo, cloth, 15s.

——————— Vol. 2. *In the press.*

Morley (Henry) — English Writers. To be completed in 3 Vols. Part I. Vol. I. **The Celts and Anglo-Saxons.** With an Introductory Sketch of the Four Periods of English Literature. Part 2. **From the Conquest to Chaucer.** (Making 2 vols.) 8vo, cloth, 22s.

*** Each part is indexed separately. The two Parts complete the account of English Literature during the Period of the Formation of the Language, or of THE WRITERS BEFORE CHAUCER.

Morley (Henry)—English Writers. Vol. II. Part I.
From Chaucer to Dunbar. 8vo, cloth, 12s.

—— **Tables of English Literature.** Containing 20 Charts.
Second Edition, with Index. Royal 4to, cloth, 12s.
In Three Parts. Parts I. and II., containing Three Charts, each 1s. 6d.
Part III. containing 14 Charts, 7s. Part III. also kept in Sections 1, 2, and 5, 1s. 6d. each, 3 and 4 together, 5s. *** The Charts sold separately.

—— **Clement Marot, and other Studies.** 2 vols.
Post 8vo, cloth, 18s.

Morley (John)—Critical Miscellanies. 8vo, cloth, 14s.

————— **Voltaire.** 8vo, cloth, 14s.

Napier (C. O. Groom)—Tommy Try, and What He Did in Science. A Book for Boys. With 46 Illustrations. Crown 8vo, 6s.

Napier (Maj.-Gen. W. C. E.)—Outpost Duty. By General Jarry, translated with Treatises on Military Reconnaissance and on Road-Making. Second Edition. Crown 8vo, 5s.

Norway: its People, Products, and Institutions. By Rev. J. Bowden. Crown 8vo, 7s. 6d.

O'Neil (Henry)—Two Thousand Years Hence. With Frontispiece and Vignette by J. Gilbert. Crown 8vo, 9s.

—— **Satirical Dialogues.** 12mo, cloth, 2s. 6d.

—— **The Age of Stucco, a Satire.** Post 8vo, cloth, 5s.

——· **Modern Art in England and France.** Post 8vo, sewed, 1s.

OUIDA'S NOVELS.

Folle-Farine. A Novel. 3 Vols. Post 8vo, cloth, 31s. 6d.

Cheap Editions.

Idalia. Crown 8vo, 5s.

Chandos. Crown 8vo, 5s.

Under Two Flags. Crown 8vo, 5s.

Cecil Castlemaine's Gage. Crown 8vo, 5s.

Tricotrin; the Story of a Waif and Stray. Crown 8vo, 5s.

Strathmore, or Wrought by his Own Hand. Crown 8vo, 5s.

Held in Bondage, or Granville de Vigne. Crown 8vo, 5s.

Puck. His Vicissitudes, Adventures, &c. Crown 8vo, 5s.

Our Farm of Four Acres. How we Managed it, the Money we Made by it, and How it Grew to one of Six Acres. Thirteenth edition, Enlarged and Illustrated. Post 8vo, cloth, 2s. 6d.

Our Life in Japan. By R. Mountency Jephson, and F. Pennell Elmhirst, 9th Regt. Demy 8vo, with numerous Illustrations from Photographs by Lord Walter Kerr, Signor Beato, and native Japanese Drawings, 18s.

Puckett (R. Campbell, Head Master of the Bath School of Art) —**Sciography**; or Radial Projection of Shadows. New edition. Crown 8vo, cloth, 6s.

Puck; His Vicissitudes, Adventures, &c. By OUIDA. Crown 8vo, 5s.

Reclus (Élisée)—**The Earth.** A Descriptive History of the Phenomena of the Life of the Globe. Section 1 and 2, Continents. Translated by the late B. H. Woodward, M.A., and Edited by Henry Woodward, British Museum. Illustrated by 250 Maps inserted in the text, and 21 page Maps printed in Colours. 2 vols., 8vo, cloth, 26s.

Raleigh, Life of Sir Walter, 1552—1618. By J. A. St. John. New edition, post 8vo, 10s. 6d.

Recollections of Eton. By an Etonian. Illustrated by Sidney P. Hall. Crown 8vo, cloth gilt, 12s.

Redgrave (Richard)—**Manual and Catechism on Colour.** 24mo, cloth, 9d.

Reynolds (Rev. R. Vincent)—**The Church and the People**; or, The adaptation of the Church's Machinery to the Exigencies of the Times. Post 8vo, 6s.

Ridge (Dr. Benjamin)—**Ourselves, Our Food, and Our Physic.** Eleventh edition, fcap 8vo, cloth, 1s. 6d.

Rimmel (Eugène)—**The Book of Perfumes.** Fourth edition, with Two Hundred Illustrations, post 8vo, cloth, 5s.

——— **Le Livre des Parfums.** Préface par Alphonse Karr, with many coloured Illustrations, 8vo, cloth, gilt, 8s.

Roba di Roma. By W. W. STORY. Sixth Edition, with Additions, and Portrait. In one vol. 10s. 6d.

Roberts (Sir Randal, Bart.)—**Glenmáhra**; or the Western Highlands, with illustrations. Crown 8vo, 6s.

——— **Modern War**; or the Campaign of the First Prussian Army, 1870—1871. With Map, 8vo, cloth, 14s.

Robinson (J. C.)—**Italian Sculpture of the Middle Ages and Period of the Revival of Art.** A Descriptive Catalogue of that section of the South Kensington Museum comprising an Account of the Acquisitions from the Gigli and Campana Collections. With Twenty Engravings. Royal 8vo, cloth, 7s. 6d.

Rock (Dr.) On Textile Fabrics. A Descriptive Catalogue of the Collection of Church Vestments, Dresses, Silk Stuffs, Needlework and Tapestries in the South Kensington Museum. By the VERY REV. CANON ROCK, D.D. Royal 8vo, half morocco, 31s. 6d.

Sarcey (Francisque).—Paris during the Siege. Translated from the French. With a Map. Post 8vo, cloth, 6s. 6d.

Sciography; or, Radial Projection of Shadows. By R. CAMPBELL PUCKETT, Ph.D., Head Master of the Bath School of Art. New Edition. Crown 8vo, 6s.

Seddall (Rev. Henry)—Malta: Past and Present; a History of Malta from the days of the Phœnicians to the present time. With Map. Demy 8vo, cloth, 12s.

Shaftesbury (Earl of)—Speeches upon Subjects having relation chiefly to the Claims and Interests of the Labouring Class. With a Preface. Crown 8vo, 8s.

Shakespeare (Dyce's). New edition, in Nine Volumes, demy 8vo, —**The Works of Shakespeare.** Edited by the Rev. Alexander Dyce. This edition is not a mere reprint of that which appeared in 1857, but presents a text very materially altered and amended from beginning to end, with a large body of critical Notes almost entirely new, and a Glossary, in which the language of the poet, his allusions to customs, &c., are fully explained. 9 vols, demy 8vo, cloth, 4l. 4s.

"The best text of Shakespeare which has yet appeared........Mr. Dyce's Edition is a great work, worthy of his reputation, and for the present it constitutes the standard text."—*Times.*

Shakespeare, Dyce's Glossary to the Works of. Demy 8vo, 12s.   This forms Vol. 9 of 'The Works.'

Simonin, L.—Underground Life; or Mines and Miners. Translated, Adapted to the Present State of British Mining, and Edited by H. W. Bristowe, F.R.S., of the Geological Survey, &c. With 160 Engravings on Wood, 20 Maps Geologically coloured, and 10 Plates of Metals and Minerals printed in Chromo-lithography. Imperial 8vo. Roxburghe binding, 42s.

Smith (Albert)—Wild Oats and Dead Leaves. Second edition, crown 8vo, cloth, 5s.

Sporting Incidents in the Life of another Tom Smith. With Illustrations. Post 8vo, cloth, 8s. 6d.

Sterne's (Laurence) Life. By PERCY FITZGERALD. Illustrations. 2 vols. Post 8vo, cloth, 24s.

Story (The) of the Commune, by a Communalist. Reprinted from the "Pall Mall Gazette," with Additions. 8vo, sewed, 1s.

Story (W. W.)—Roba di Roma. Sixth Edition, with Additions and Portrait. Post 8vo, cloth, 10s. 6d.

———— **The Proportions of the Human Frame, according to a New Canon.** With Plates. Royal 8vo, cloth, 10s.

Studies in Conduct. Short Essays from the "Saturday Review."
Post 8vo, cloth, 7s. 6d.

Tainsh (E. C.)—A Study of the Works of Alfred Tennyson, D.C.L., Poet Laureate. New Edition, with Supplementary Chapter on the "HOLY GRAIL." Crown 8vo, cloth, 6s.

Townshend (Chauncy Hare) — The Religious Opinions of the late Rev. Chauncy Hare Townshend. Published as directed in his Will, by his Literary Executor. Crown 8vo, 9s.

Trinal. Memorials of Theophilus Trinal, Student. By the Rev. T. T. LYNCH. *New edition, enlarged.* Crown 8vo, cloth extra, 6s.

TROLLOPE'S (ANTHONY) WORKS.

Can You Forgive Her? With Forty Illustrations, 8vo, cloth, 7s. 6d.
—— Cheap Edition. 12mo, boards, 3s.; cloth, 4s.
Orley Farm. With Forty Illustrations by J. E. MILLAIS. 8vo, cloth, 7s. 6d.
—— Cheap Edition, 12mo, boards, 3s.; cloth, 4s.
Hunting Sketches. Second Edition, post 8vo, cloth, 3s. 6d.
Travelling Sketches. Second Edition, post 8vo, cloth, 3s. 6d.
Clergymen of the Church of England. Post 8vo, cloth, 3s. 6d.
Dr. Thorne. 12mo, boards, 2s.; cloth, 2s. 6d.
The Bertrams. 12mo, boards, 2s.; cloth, 2s. 6d.
The Kellys and the O'Kellys. 12mo, bds. 2s.; cloth, 2s. 6d.
The Macdermots of Ballycloran. 12mo, boards, 2s.; cloth, 2s. 6d.
Castle Richmond. 12mo, boards, 2s.; cloth, 2s. 6d.
Miss Mackenzie. 12mo, boards, 2s.; cloth, 2s. 6d.
Rachel Ray. 12mo, boards, 2s.; cloth, 2s. 6d.
Tales of all Countries. 12mo, boards, 2s.; cloth, 2s. 6d.
The Belton Estate. 12mo, boards, 2s.; cloth, 2s. 6d.
—— Post 8vo, cloth, 5s.
Phineas Finn, the Irish Member. 12mo, boards, 3s.
He knew He was Right. 12mo, boards, 3s.; cloth, 4s.
Mary Gresley. 12mo, boards, 2s.; cloth, 2s. 6d.
Lotta Schmidt. 12mo, boards, 2s.; cloth, 2s. 6d.

Trollope (Thomas Adolphus) — A History of the Commonwealth of Florence. From the Earliest Independence of the Commune to the Fall of the Republic in 1531. 4 vols., demy 8vo, cloth, 3l.

Turnor (Hatton)—Astra Casta. Experiments and Adventures in the Atmosphere. With upwards of 100 Engravings and Photozincographic Plates produced under the superintendence of COLONEL SIR HENRY JAMES, R.E. Second Edition, royal 4to, cloth, 35s.

Underground Life; or, Mines and Miners. By L. SIMONIN. Translated, Adapted to the Present State of British Mining, and Edited by H. W. Bristowe, F.R.S., of the Geological Survey, &c. Imperial 8vo, with 160 Engravings on Wood, 20 Maps Geologically Coloured, and 10 Plates of Metals and Minerals printed in Chromo-lithography. Roxburghe binding. 42s.

Universal Catalogue of Books on Art. Compiled for the use of the National Art Library, and the Schools of Art in the United Kingdom. In 2 vols., crown 4to, half morocco, 21s. each.

Verne (Jules)— Five Weeks in a Balloon. A Voyage of Exploration and Discovery in Central Africa. Translated from the French. With 64 Illustrations, post 8vo, 7s. 6d.

Vesinier's, P., (*Ex-member and Secretary of the Commune, and Rédacteur en chef du Journal Officiel*) **History of the Commune of Paris.** Post 8vo, cloth, 7s. 6d.

Voltaire. By John Morley, *Editor of the Fortnightly Review.* 8vo, cloth, 14s.

Weber (Carl Maria von)—The Life of an Artist. By HIS SON. Translated by J. P. SIMPSON. 2 vols., post 8vo, cloth, 22s.

Whist Player (The). By Colonel Blyth. With Coloured Plates of 'Hands.' Third edition, imperial 16mo, cloth, 5s.

White (Walter)—Eastern England. From the Thames to the Humber. 2 vols., post 8vo, cloth, 18s.

—— **Month in Yorkshire.** Fourth Edition, with a Map, post 8vo, cloth, 4s.

—— **Londoner's Walk to the Land's End, and a Trip to the Scilly Isles.** With Four Maps. Second Edition, post 8vo, 4s.

Wornum (R. N.)—The Epochs of Painting. A Biographical and Critical Essay on Painting and Painters of all Times and many Places. With numerous Illustrations. Demy 8vo, cloth, 20s.

—— **Analysis of Ornament — The Characteristics of Styles.** An Introduction to the Study of the History of Ornamental Art. With many Illustrations. Second Edition. Royal 8vo, cloth, 8s.

Wornum (R. N.)—Some Account of the Life of Holbein, Painter of Augsburg. With Portrait and 34 Illustrations. Imperial 8vo, cloth, 31s. 6d.

Wynter (Dr.)—Curiosities of Toil, and Other Papers. 2 vols., post 8vo, 18s.

BOOKS FOR THE USE OF SCHOOLS.

Issued under the Authority of the Science and Art Department, South Kensington.

An Alphabet of Colour. Reduced from the works of FIELD, HAY, CHEVREUIL. 4to, sewed, 3s.

Art Directory. 12mo, sewed, 6d. [*Annual.*]

Bradley (Thomas), of the Royal Military Academy, Woolwich—Elements of Geometrical Drawing. In Two Parts, with Sixty Plates, oblong folio, half-bound, each part, 16s.

—— Selection (from the above) of Twenty Plates, for the use of the Royal Military Academy, Woolwich. Oblong folio, half-bound, 16s.

Burchett's Linear Perspective. With Illustrations. Post 8vo, cloth, 7s.

—— Practical Geometry. Post 8vo, cloth, 5s.

—— Definitions of Geometry. Third Edition, 24mo, sewed, 5d.

Davidson (Ellis A.) — Drawing for Elementary Schools. Post 8vo, cloth, 3s.

—— Orthographic and Isometrical Projection. 12mo, cloth, 2s.

—— Linear Drawing. Geometry applied to Trade and Manufactures. 12mo, cloth, 2s.

—— Drawing for Carpenters and Joiners. 12mo, cloth, 3s. 6d.

—— Building Construction and Architectural Drawing. 12mo, cloth, 2s.

—— Model Drawing. 12mo, cloth, 3s.

—— Practical Perspective. 12mo, cloth, 3s.

Delamotte (P. H.)—Progressive Drawing Book for Beginners. 12mo, 2s. 6d.

Dicksee (J. R.)—School Perspective. 8vo, cloth, 5s.

Directions for Introducing Elementary Drawing in Schools and among Workmen. Published at the Request of the Society of Arts. Small 4to, cloth, 4s. 6d.

Drawing for Young Children, 150 Copies. 16mo, cloth, 3s. 6d.

Dyce's Drawing Book of the Government School of Design, Elementary Outlines of Ornament. 50 Plates, small folio, sewed, 5s.

——— Introduction to ditto. Foolscap 8vo, 6d.

Educational Division of S. K. Museum. Classified Catalogue of, 8vo, reprinting.

Elementary Drawing Copy-Books, for the use of Children from four years old and upwards, in Schools and Families. Compiled by a Student certificated by the Science and Art Department as an Art Teacher. Seven Books in 4to, sewed:—

Book I. Letters, 8d.
" II. Ditto, 8d.
" III. Geometrical and Ornamental Forms, 8d.
" IV. Objects, 8d.
" V. Leaves, 8d.
" VI. Birds, Animals, &c., 8d.
" VII. Leaves, Flowers, and Sprays, 8d.

. Or in sets of Seven Books, 4s. 6d.

Engineer and Machinist Drawing Book, 16 parts, 71 plates, folio, 32s.

ditto " " ditto " 15 by 12in., mounted, 64s.

Examination papers for Science Schools and Classes. [Annual.]

Foster (Vere)—Drawing Copy Books. Fcap 4to, 1d. each.

ditto " " fine paper with additions, fcap 4to, 3d. each.

Gregory (Chas.)—First Grade Freehand Outline Drawing Examples (for the black board), 4to, packet, 2s. 6d.

Henslow (Prof.)—Illustrations to be employed in the Practical Lessons on Botany. Prepared for South Kensington Museum. Post 8vo, sewed, 6d.

Hulme (F. E.)—Sixty Outline Examples of Freehand Ornament, royal 8vo, half bound, 10s. 6d.; or, in six parts, each, 1s. 6d.

Jewitt's Handbook of Practical Perspective. 18mo, cloth, 1s. 6d.

Kennedy (John)—First Grade Practical Geometry, 12mo, 6d.

———— **Freehand Drawing Book,** 16mo, cloth, 1s. 6d.

Laxton's Examples of Building Construction, 1 and 2 divisions, folio, each containing 16 plates, 10s. each.

Lindley (John)—Symmetry of Vegetation, principles to be observed in the delineation of plants. 12mo, sewed, 1s.

Marshall's Human Body. Text and Plates, 2 vols., cloth, 21s.

Principles of Decorative Art. Folio, sewed, 1s.

Puckett, R. Campbell (head master of the Bath School of Art)— **Sciography or Radial Projection of Shadows,** crown 8vo, cloth, 6s.

Redgrave's Manual and Catechism on Colour. Second edition, 24mo, sewed, 9d.

Robinson's (J. C.) Lecture on the Museum of Ornamental Art, Fcap. 8vo, sewed, 6d.

———— **Manual of Elementary Outline Drawing for the Course of Flat Examples,** 32mo, 7d.

Science Directory, 12mo, sewed, 6d. [*Annual.*]

Wallis (George)—Drawing Book, oblong, sewed, 3s. 6d.

    ditto   ,,   ditto, mounted, 8s.

Wornum (R. N.)—The Characteristics of Styles; An Introduction to the Study of the History of Ornamental Art, royal 8vo, cloth, 8s.

Wornum (R. N.)—Catalogue of Ornamental Casts, 8vo, cloth, 1s. 6d.

Outline Examples.

A. C. &. Letters, 3 sheets, 1s., mounted, 3s.
Albertolli, Selections of Foliage from, 4 plates, 3d., mounted, 3s. 6d.
Familiar Objects. Mounted, 9d.
Flowers Outlined from the Flat. 8 sheets, 8d., mounted, 2s. 6d.
Morghen's Outline of Human Figure, by Herman, 20 sheets, 3s. 4d., mounted, 10s.
Simpson's 12 Outlines for Pencil Drawing, mounted, 7s.
Tarsia. Ornament Outlined from the Flat. Wood Mosaic, 4 plates, 7d., mounted, 3s. 6d.
Trajan Frieze from the Forum of Trajan. Part of a, 4d., mounted, 1s.
Waithricht's Outlines of Ornament, by Herman, 12 sheets, 2s., mounted, 5s. 6d.
Delarue's Flat Examples for Drawing, Objects, 48 subjects, in packet, 3s.
—— —— Animals, in packet, 1s.
Dyce's Elementary Outlines of Ornament. Drawing Book of the Government School of Design, 50 plates, sewed, 5s., mounted, 18s.
—— Selection of 15 plates from do., mounted, 6s. 6d.
Smith's (W.) Examples of First Practice in Freehand Outline Drawing, oblong, sewed, 2s.
Wallis's Drawing Book, oblong, sewed, 3s. 6d., mounted, 8s.

Shaded Examples.

Bargue's Course of Design, 20 selected sheets, each sheet, 2s. 3d.
Doric Renaissance Frieze Ornament (shaded ornament), sheet 4d., mounted, 1s. 2d.
Early English Capital, sheet, 4d., mounted, 1s.
Gothic Patera, sheet, 4d., mounted, 1s.
Greek Frieze, From a, sheet, 3d., mounted, 9d.
Pilaster, Part of a, from the tomb of St. Biagio, at Pisa, sheet 1s., mounted, 2s.
Renaissance Scroll, sheet, 1s. 4d., mounted, 2s.
Renaissance Rosette, sheet, 3d., mounted, 6d.
Sculptured Foliage, Decorated, Moulding of, sheet, 7d., mounted, 1s. 2d.
Smith's Diagrams for the Black Board, 10 in packets, 2s.
Column from the Vatican, sheet, 1s., mounted, 2s.
White Grapes, sheet, 9d., mounted, 2s.
Virginia Creeper, sheet, 9d., mounted, 2s.
Burdock, sheet, 4d., mounted, 1s. 2d.
Poppy, sheet, 4d., mounted, 1s. 2d.
Foliated Scroll from the Vatican, sheet, 5d., mounted, 1s. 3d.

Coloured Examples.

Camellia, sheet, 2s. 9d., mounted, 3s. 9d.
Pelargonium, sheet, 2s. 9d., mounted, 3s. 9d.
Petunia, sheet, 2s. 9d., mounted, 3s. 9d.
Nasturtium, sheet, 2s. 9d., mounted, 3s. 9d.
Oleander, sheet, 2s. 9d., mounted, 3s. 9d.
Group of Camellias, mounted, 12s.
Diagram to illustrate the Harmonious Relations of Colour, sheet, 9d., mounted, 1s. 6d.
Elementary Design, 2 plates, sheet, 1s.
Pyne's Landscapes in Chromolithography, (six) each, mounted, 7s. 6l.
Cotman's Pencil Landscapes, (nine) set, mounted, 15s.
—— Sepia —— (five) set, mounted, 20s.
Downs Castle, Chromolithograph, mounted, 7s.

Petit (Stanislas)—Selected Examples of Machines of Iron and Woodwork (French), 60 sheets, each 1s. 1d.

Tripon (J. B.)—Architectural Studies, 20 plates, each, 1s. 8d.

Lineal Drawing Copies, in portfolio, 5s. 6d.

Design of an Axminster Carpet, by MARY JULYAN. 2s.

MODELS AND INSTRUMENTS.

A Box of Models for Parochial Schools, 1l. 4s.

Binn's Box of Models for Orthographic Projection applied to Mechanical Drawing, in a box, 30s.

Davidson's Box of Drawing Models, 40s.

Rigg's Large (Wood) Compasses, with Chalk Holder. 4s. 3d.

Set of Large Models. A Wire Quadrangle, with a Circle and Cross within It, and one Straight Wire. A Solid Cube. A Skeleton Wire Cube. A Sphere. A Cone. A Cylinder. A Hexagonal Prism, 2l. 2s.

Models of Building Construction. Details of a king-post truss, £2.

———— Details of a six-inch trussed partition for floor, £3 3s.

———— Details of a trussed timber beam for a traveller, £4 10s.

These models are constructed in wood and iron.

Skeleton Cube in Wood, 3s. 6d.

A Stand with a Universal Joint, to show the Solid Models, &c., 1l. 10s.

Slip, Two set squares, and T-square, 5s.

Specimens of the Drawing-board, T-square, Compasses, Books on Geometry and Colour, Case of Pencils and Colour-box; awarded to Students in Parish Schools, 13s. 6d.

Imperial Deal Frames, glazed, without sunk rings, 10s.

Elliott's Case of Instruments, containing 6-in. compasses with pen and pencil leg, 6s. 9d.

———— Prize Instrument Case, with 6-in. compasses, pen and pencil leg, two small compasses, pen and scale, 18s.

———— 6-in. Compasses, with shifting pen and point, 4s.

MODELS AND INSTRUMENTS—continued.

Three Objects of *Form* **in Pottery (Minton's)—Indian Jar; Celadon Jar; Bottle,** 13s. 9d.

Five selected Vases in Majolica Ware (Minton's), 2l. 2s. 6d.

Three selected Vases in Earthenware (Wedgwood's), 15s. 6d.

LARGE DIAGRAMS.

Astronomical. *Twelve sheets.* Prepared for the Committee of Council of Education by JOHN DREW, Ph.Dr., F.R.S.A., *each sheet,* 4s.

——— on rollers and varnished, *each,* 7s.

Building Construction. By WILLIAM J. GLENNY, Professor of Drawing, King's College. 10 sheets. In sets, 21s.

Physiological. *Nine sheets.* Illustrating Human Physiology, Life-size and Coloured from Nature, prepared under the direction of JOHN MARSHALL, M.R.C.S., *each sheet,* 12s. 6d.

1. SKELETON AND LIGAMENTS.
2. MUSCLES, JOINTS, &c.
3. VISCERA AND LUNGS.
4. HEART AND BLOOD VESSELS.
5. LYMPHATICS OR ABSORBENTS.
6. DIGESTIVE ORGANS.
7. BRAIN AND NERVES.
8. ORGANS OF THE SENSES.
9. TEXTURES, MICROSCOPIC STRUCTURE.

On canvas and rollers, varnished, *each,* 21s.

Zoological. *Ten sheets.* Illustrating the Classification of Animals, by ROBERT PATTERSON, *each sheet,* 4s.

——— on canvas and rollers, varnished, *each,* 7s.

The same, reduced in size, on Royal paper, *in nine sheets,* 12s.

Botanical. *Nine sheets.* Illustrating a Practical Method of Teaching Botany, by Professor HENSLOW, F.L.S., 40s.

——— on canvas and rollers, and varnished, 3l. 3s.

Extinct Animals. *Six sheets.* By B. WATERHOUSE HAWKINS, F.C.S., in tinted Lithography, on canvas and rollers, and varnished, *each,* 8s. 10d.

Mechanical. *Six sheets.* Pump, Hydraulic Press, Water Wheel, Turbine, Locomotive Engine, Stationary Engine, 62½-in. by 47-in., on canvas and roller, *each* 16s. 6d.

Illustrations of the principal Natural Orders of the Vegetable Kingdom. By Professor OLIVER, F.R.S., F.L.S. Seventy Imperial sheets containing examples of dried plants, representing the different orders. Five guineas the set.

THE
FORTNIGHTLY REVIEW.

Edited by JOHN MORLEY.

THE object of THE FORTNIGHTLY REVIEW is to become an organ for the unbiassed expression of many and various minds on topics of general interest in Politics, Literature, Philosophy, Science, and Art. Each contribution will have the gravity of an avowed responsibility. Each contributor, in giving his name, not only gives an earnest of his sincerity, but is allowed the privilege of perfect freedom of opinion, unbiassed by the opinions of the Editor or of fellow-contributors.

THE FORTNIGHTLY REVIEW is published on the 1st of every month (the issue on the 15th being suspended), price Two Shillings, and a Volume is completed every Six Months.

The following are among the Contributors:—

J. S. MILL.	PROFESSOR BEESLY.	W. T. THORNTON.
PROFESSOR HUXLEY.	A. C. SWINBURNE.	PROFESSOR BAIN.
PROFESSOR TYNDALL.	DANTE GABRIEL ROSSETTI.	PROFESSOR FAWCETT.
DR. VON SYBEL.	HERMAN MERIVALE.	HON. R. LYTTON.
PROFESSOR CAIRNES.	EDWARD A. FREEMAN.	ANTHONY TROLLOPE.
EMILE DE LAVELEYE.	WILLIAM MORRIS.	JOSEPH MAZZINI.
GEORGE HENRY LEWES.	F. W. FARRAR.	THE EDITOR.
FREDERIC HARRISON.	PROFESSOR HENRY MORLEY.	
WALTER BAGEHOT.	J. HUTCHISON STIRLING.	&c. &c. &c.

Contents for November.
SECOND EDITION.
CONTAINS:
JOHN STUART MILL ON BERKELEY'S LIFE AND WRITINGS.
PROFESSOR HUXLEY ON ADMINISTRATIVE NIHILISM.
HENRY FAWCETT, M.P., ON THE PRESENT POSITION OF THE GOVERNMENT.
WALTER H. PATER ON THE POETRY OF MICHAEL ANGELO.
JULES ANDRIEU ON THE PARIS COMMUNE; A CHAPTER TOWARDS ITS THEORY AND HISTORY.
ANTHONY TROLLOPE'S THE EUSTACE DIAMONDS.

Contents for December.
SPECIALIZED ADMINISTRATION. By HERBERT SPENCER.
CHURCH AND STATE IN ITALY. By J. W. PROBYN.
THE EUSTACE DIAMONDS. Chapters XXI. to XXIV. By ANTHONY TROLLOPE.
PHYSICS AND POLITICS. IV. NATION-MAKING. By WALTER BAGEHOT.
THE NEW ATTACK ON TOLERATION. By HELEN TAYLOR.
LYRICAL FABLES (Continued). By the Hon. ROBERT LYTTON.
THE IRISH UNIVERSITY QUESTION. By H. DIX HUTTON.

Contents for January.
THE POSITION AND PRACTICE OF THE HOUSE OF LORDS. By LORD HOUGHTON.
THE CLOUD CONFINES. By DANTE GABRIEL ROSSETTI.
HOME RULE. By W. O'CONNOR MORRIS.
CHAUMETTE. By A. BESSARD.
PHYSICS AND POLITICS. V. THE AGE OF DISCUSSION. By WALTER BAGEHOT.
NEW THEORIES IN POLITICAL ECONOMY. By PROFESSOR CAIRNES.
ST. BERNARD OF CLAIRVAUX. By J. C. MORISON.
THE EUSTACE DIAMONDS. Chapters XXV. to XXVIII. By ANTHONY TROLLOPE.
FORSTER'S LIFE OF DICKENS.

CHAPMAN & HALL, 193, PICCADILLY.

www.ingramcontent.com/pod-product-compliance
Lightning Source LLC
Chambersburg PA
CBHW030730230426
43667CB00007B/664